Second Edition

Safety Pharmacology in Pharmaceutical Development

Approval and Post Marketing Surveillance

Second Edition

Safety Pharmacology in Pharmaceutical Development

Approval and Post Marketing Surveillance

Shayne C. Gad

CRC Press
Taylor & Francis Group
Boca Raton London New York

CRC Press is an imprint of the
Taylor & Francis Group, an **informa** business

CRC Press
Taylor & Francis Group
6000 Broken Sound Parkway NW, Suite 300
Boca Raton, FL 33487-2742

First issued in paperback 2019

© 2012 by Taylor & Francis Group, LLC
CRC Press is an imprint of Taylor & Francis Group, an Informa business

No claim to original U.S. Government works

ISBN-13: 978-1-4398-4567-7 (hbk)
ISBN-13: 978-0-367-38145-5 (pbk)

Library of Congress Cataloging-in-Publication Data

Gad, Shayne C., 1948-
   Safety pharmacology in pharmaceutical development : approval and post marketing
surveillance / Shayne C. Gad. -- 2nd ed.
      p. ; cm.
   Rev. ed. of: Safety pharmacology in pharmaceutical development and approval. c2004.
   Includes bibliographical references and index.
   ISBN 978-1-4398-4567-7 (hardback : alk. paper)
   I. Gad, Shayne C., 1948- Safety pharmacology in pharmaceutical development and approval.
II. Title.
   [DNLM: 1. Drug Evaluation, Preclinical. 2. Drug Approval. 3. Product Surveillance,
Postmarketing. 4. Toxicity Tests. QV 771]

   615.7'04--dc23   2012009446

Visit the Taylor & Francis Web site at
http://www.taylorandfrancis.com

and the CRC Press Web site at
http://www.crcpress.com

Dedication

To Novie Beth—
for bringing so much
light and joy
to this life

Contents

Preface

Safety pharmacology is the evaluation and study of the pharmacological effects of a potential drug that are unrelated to the desired therapeutic effect and, therefore, may present a hazard—particularly in individuals who already have one or more compromised or limited organ system functions, as is generally the case for those receiving pharmacotherapy. Unlike other nonclinical evaluations of a drug's safety, these safety pharmacology evaluations are usually conducted at doses close to the intended clinical dose. Continued International Conference on Harmonisation of Technical Requirements for Registration of Pharmaceuticals for Human Use (ICH) guidelines, followed by versions by national regulatory authorities including those of the U.S. Food and Drug Administration (FDA), have made such evaluations mandatory before a potential drug is introduced into humans—while also failing to provide clear guidance as to how the requirements are to be met.

General/safety pharmacology has been an emerging discipline within the pharmaceutical industry over the last 30 years in which unanticipated effects of new drug candidates on major organ function (i.e., secondary pharmacological effects) are critically assessed in a variety of animal models. A survey was conducted to obtain customer input on the role and strategies of this emerging discipline. Overlooked in importance by all but a few (Zbinden, 1966, 1984) for many years, the Japanese clearly became the leaders in developing and requiring such information, while the United States was (and remains) in a position behind Japan and the European Union (EU) in establishing formal requirements and in implementing industrial programs.

Most companies have traditionally conducted evaluations of cardiovascular and central nervous system (CNS) functions, while few have conducted respiratory, gastrointestinal, and renal function evaluations; a few conduct a ligand-binding/activity panel as part of their pharmacological profiling. Resources needed to complete a company's standard safety pharmacology program are approximately one to four full-time people per compound. One third of companies use a maximum tolerated dose (MTD) for safety pharmacology studies; two thirds use multiples of pharmacological or therapeutic doses. Until 2002, only half conducted safety pharmacology studies with Good Laboratory Practices (GLPs), using the 1992 Japanese guidelines as a guide or outline. Company clinicians were most often cited as the primary customers for whom safety pharmacology studies were done, followed by research and development scientists and then regulatory authorities. These results suggest that most companies primarily conducted safety pharmacology for its contribution to risk assessment and critical care management.

It is important that the tests employed detect bidirectional drug effects and be validated in both directions with appropriate reference (control) substances. This requirement is less appropriate for multiparameter procedures. Blind testing could be an advantage. Ethical considerations are important, but the ultimate ethical criterion is the assessment of risk for humans. Safety pharmacology studies should not be overly inclusive but should be performed to the most exacting standards, including GLP compliance. This is, of course, backward; such human tolerance is properly an extension (and expression) of the nonclinical safety pharmacology.

The other point of view has been that properly executed, repeated-dose preclinical safety studies meeting the current design theoretically could fill these needs, recognizing that undesired pharmacological activities of novel drugs or biologicals may limit development of a therapeutic agent prior to the characterization of any toxicological effects. In rodent species, general pharmacology assays have traditionally been used to screen new agents for pharmacological effects on the central and peripheral nervous systems, the autonomic nervous system and smooth muscles, the respiratory and cardiovascular systems, the digestive system, and the physiological mechanisms of water and electrolyte balance. In large animal species, such as dogs and nonhuman primates, smaller numbers of animals per study limit their use for screening assays, but these species may play an important role in more detailed mechanistic studies. For drugs and biologicals that must be tested in nonhuman primates because of species-specific action of the test agent, functional pharmacology data are often collected during acute or subacute toxicity studies. This requires careful experimental design to minimize any impact that pharmacological effects or instrumentation may have on the assessment of toxicity. In addition, with many new therapies targeted at immunological diseases, the pharmacological effect of therapeutics on the immune system presents new challenges for pharmacology profiling and has now been addressed by its own ICH guidance (ICH S8). The applications of pharmacology assay by organ system in both rodent and large animal species are discussed, as well as practical issues in assessing pharmacological end points in the context of toxicity studies (Martin et al. 1997; Matsuzawa et al. 1997).

In Europe, the numbers of registered drugs and drug expenditure are increasing rapidly. Within the EU, regulations no longer require that new drugs have to be better than old ones. At the same time, pharmacoepide-miology studies in Europe and the United States continue to show that adverse drug reactions now may account for up to 10% of the admissions of patients to hospitals at a cost of hundreds of millions of U.S. dollars annually (Sjouist, 2000). Compared with 30 years ago, this represents a considerable increase. The many shortcomings of clinical trials and their relevance to health care provide a partial explanation. Adverse drug reactions are often poorly studied and documented in these trials and seldom included in health economical analyses of the value of new drugs. Pharmacovigilance is product oriented rather than utilization oriented and quite invisible in clinical medicine. This is regrettable because up to 50% of adverse drug reactions (ADRs) are dose dependent and thus preventable. Hopefully, the rapid progress in molecular and clinical pharmacogenetics will provide new tools to enable clinicians to choose and dose drugs according to the needs of individual patients. A good starting point for those not well versed in pharmacology and the range of potential mechanisms of action and interaction can be found in Goodman and Gilman's *The Pharmacological Basis of Therapeutics* (Brunton, Lazo, and Parker, eds., 11th ed., McGraw-Hill, 2006).

In this essential and rapidly changing field, it is hoped that this first volume on the subject will answer many questions and add clarity to existing requirements.

References

Martin, L.I., Horvath, C.J., and Wyand, M.S., Safety pharmacology screening: Practical problems in drug development, *Int. J. Toxicol.*, 1997, 16:41–65.

Matsuzawa, T., et al., Current status of conducting function tests in repeated dose toxicity studies in Japan, *J. Toxicol. Sci.*, 1997, 22:374–382.

Sjouist, F., Drug-related hospital admissions, *Ann. Pharmacol.*, 2000, 34:832–839.

Zbinden, G., Neglect of function and obsession with structure in toxicity testing. In: *Proc. 9th Int. Cong. Pharmacol.*, vol. 1, New York: Macmillan, 1984, pp. 43–49.

Zbinden, G., The significance of pharmacologic screening tests in the preclinical safety evaluation of new drugs, *J. New Drugs*, 1966, 6:1–7.

About the Author

Shayne C. Gad, B.S. (Whittier College, California, chemistry and biology, 1970) and Ph.D. in pharmacology/toxicology (University of Texas at Austin, 1977), Diplomate of the American Board of Toxicologists, Academy of Toxicological Sciences, is the principal of Gad Consulting Services, a 19-year-old consulting firm with six employees and more than 500 clients (including 120 pharmaceutical companies in the United States and 50 overseas). Prior to this, he served in director-level and above positions at Searle, Synergen, and Becton Dickinson. He has authored or edited more than 44 published books and more than 360 chapters, articles, and abstracts in the fields of toxicology, statistics, pharmacology, drug and medical device development, and safety assessment. He has more than 34 years of broad-based experience in toxicology, drug and device development, statistics, and risk assessment. He has specific expertise in neurotoxicology, *in vitro* methods, cardiovascular toxicology, inhalation toxicology, immunotoxicology, and genotoxicology. He is a past president of the American College of Toxicology, the Roundtable of Toxicology Consultants, and three of Society of Toxicology's specialty sections and is a recipient of the American College of Toxicology Lifetime Contribution Award. He has had direct involvement in the preparation of investigational device exemptions (INDs) (95 successfully to date), new drug applications (NDAs), product license applications (PLAs), abbreviated new drug applications (ANDAs), 510(k), investigational device exemptions (IDEs), comparative toxicogenomics database (CTD), product marketing applications (PMAs) and clinical data bases for Phase 1 and 2 studies. He has consulted for the FDA, U.S. Environmental Protection Agency (EPA), and National Institutes of Health (NIH) and has trained reviewers and been an expert witness for the FDA. He has also conducted the triennial toxicology salary survey as a service to the profession for the past 22 years.

chapter one

Safety pharmacology

Background, history, issues, and concerns

Although our current approach to nonclinical evaluation of the safety of drugs generally serves us well in identifying potential adverse effects of drugs before they go into man, almost all drugs in clinical use have side effects, serious and not so serious, that are only identified after they are in clinical use.

> The adverse drug reactions which the standard toxicological test procedures do not aspire to recognize include most of the functional side-effects. Clinical experience indicates, however, that these are much more frequent than the toxic reactions due to morphological and biochemical lesions [1].

Safety pharmacology is the evaluation and study of the pharmacologic effects of a potential drug (or excipient) that are unrelated to the desired therapeutic effect and therefore may present a hazard—particularly in individuals who already have one or more compromised or limited organ system functions. Unlike other nonclinical evaluations of the safety of a drug, these evaluations are usually conducted at doses close to the intended clinical dose.

Such pharmacologically based adverse effects have not figured in Phase I (first in man) volunteer deaths for some years. While such volunteer deaths are very rare (only two have been reported in the United States since 1990), severe adverse events are not.

In the United States, from 1954 until 1980, 7000 healthy volunteers participated in Phase I studies with only one death (in 1980, not due to adherence to study participation) [2]. Additionally, in the United States there was one other reported volunteer death in 2001 [3].

In a typical 12-month period, surveyed in Great Britain, there were no deaths in 8163 healthy volunteers, with the three reported severe adverse events being a severe dermatitis, an anaphylactic shock, and perforation of an ulcer [4]. And in a Phase I center in France, among 1015 healthy young

Table 1.1 Risks to Healthy Volunteers in Clinical Pharmacology Studies [6]

Year	Country	Number of Volunteers	Moderately Severe AE	Potentially Life-Threatening AE	Deaths
1965–1977	USA	29,162	58 (0.2%)	—	0
1980	USA	—	—	—	1
1983	Ireland	—	—	—	1
1984	UK	—	—	—	1
1986–1987	UK	8,162	45 (0.55%)	3 (0.04%)	0
1986–1995	France	1,015	43 (3%)	0	0
2001	USA	—	—	—	1

Note: AE = adverse effect.

volunteers, there were 34 severe adverse events but no deaths [5]. None of the drugs later withdrawn from the market for QT interval prolongation were detected in such studies.

All of this suggests that screening in normal healthy volunteers is unlikely to serve to detect functional (pharmacological) effects such as have served to remove drugs from the market (subsequent to market introduction). When conducted properly, such studies are in the wrong population (a healthy and therefore fairly insensitive one) and at too low a dose (one safe to the volunteers) to detect any but the strongest signal of concern. While subsequent stages of clinical development do evaluate drugs in patients with the intended disease state, these subjects are generally from very narrow medical profiles, uncomplicated by other common disease conditions (Table 1.1).

1.1   General versus safety pharmacology

A wide variety of general pharmacology screens have been available since the 1940s and 1950s [7,8]. The value of many of these screens has been demonstrated in a number of instances where potentially important adverse effects of a novel therapeutic were identified prior to use in clinical trials [9–13]. Such general pharmacology tests evaluate affinity for a pharmacologic target (receptor), pharmacodynamic activity, and interactions in pharmacodynamic processes [14–18]. Pharmacologic evaluations are generally characterized as primary pharmacology (intended activity at the target efficacy receptors), safety pharmacology (the subject of this book), or general pharmacology (effects at other nontarget receptors). However, with the extensive advances in biomedical research in the last two decades, especially in molecular biology, immunology, and neurobiology, many novel types of therapeutic molecules are being developed. At the same time, pharmaceutical and biotechnology companies are being pressured by wary investors, competition, and advocates for the terminally ill to introduce suitably safe therapeutics more rapidly into clinical use and the marketplace. In this atmosphere, it is incumbent upon the research scientist to carefully consider each new therapeutic under development to identify potential adverse effects and determine the optimal methods for measuring the degree of safety of a compound as much as possible prior to its use in humans. There is no simple formula or set group of safety pharmacology screens that is ideal for all kinds of therapeutic agents. Knowledge of the pharmacology of a compound and any knowledge gained from traditional toxicology studies or the incorporation of some safety pharmacology endpoints within a toxicology study can help to better determine and assess the safety of new therapeutics. Working together with regulatory agencies, clinical pharmacologists, and basic scientists, those developing compounds in industry can establish the most appropriate schedule of studies and screens to safely advance their product into the clinic. Objectives of safety pharmacology studies include seeking to:

- Identify undesirable pharmacodynamic properties
- Evaluate adverse pharmacodynamic and/or pathophysiological effects
- Investigate the mechanism of the adverse pharmacodynamic effects observed and/or suspected

Safety pharmacology is a constantly evolving discipline within the pharmaceutical industry in which unanticipated effects of new drug candidates on major organ function (i.e., secondary pharmacological effects) are critically assessed in a variety of *in vitro* and animal models. Prior to the implementation of required preclinical safety pharmacology evaluations in 2001, adverse functional effects were overlooked in importance by all but a few [19–22] for many years, with the Japanese clearly having become the leaders in developing and requiring such information [23,24], while the United States was in a position behind Japan and the European Union (EU) in both having formal requirements and in implementing industrial programs. Although major companies were aware and largely addressing the need by the mid-1990s [25–28], initial Food and Drug Administration (FDA) guidelines were not proposed and promulgated until the middle of 2001, and the Committee for Proprietary Medicinal Products (CPMP) only two years before this [29–31]. The operational issues of what is to be done, how it is to be done, and how the resulting data will be used were still being worked out in the last quarter of 2002 [32], and indeed still continue to evolve. Table 1.2 presents some examples of adverse known functional events.

Table 1.2 Examples of Functional Toxicity

Compound/ Compound Class	Adverse Effect
Extensions of Primary Pharmacological Activity	
Sympatholmimetic bronchodilators	Tachycardia, palpitations
Digitalis	Hypotension arrhythmias, ventricular tachycardia
Antihistamines	Sedation
Antihypertensives	Postural hypotension
Suxamethonium	Neuromuscular blockade (esp. Prolonged apnea)
Corticosteroids	Adrenal hypofunction
Insulin	Tachycardia
Antiarrhythmics	Arrhythmia
Nitroglycerin	Hypotension
Anticoagulants	Hemorrhage
Unrelated to Primary Pharmacological Action [21]	
β-lactam antibiotics	Convulsions
Aminoglycoside antibiotics	Neuromuscular blockade
Tricyclic antibiotics	Anticholinergic effects (dry mouth, blurred vision, hypertension) arrhythmias
Emetine	Hypotension, tachycardia, EKG abnormalities
Digitalis	Gynecomastia
Antihypertensives	Sedation
Streptozolocin	Hyperglycemia
Antihistamines	Anticholinergic effects
Sulfonamides	Diarrhea, nausea, emesis
Vancomycin	Hypotension

1.2 History

The increased interest and eventual regulatory requirements for functional safety (safety pharmacology) testing has come from the occurrence of a number of drugs having to be withdrawn due to unacceptable levels of unanticipated serious adverse events and deaths occurring after drugs have entered the marketplace, mandating drug withdrawal from the marketplace [33,34].

Table 1.3 presents a listing of drugs withdrawn from the market due to safety reasons since 1990. It should be noted that only four of these withdrawals (denoted by an asterisk) were due to adverse safety pharmacology, with three of these four being cardiovascular.

Table 1.3 Postapproval Adverse Side Effects and
Related Drug Withdrawals Since 1990 [33]

• 51% of approval drugs had serious postapproval-identified side effects

Year	Drug	Indication/Class	Causative Effects
1991	Enkaid (4 years on market)	Antiarrhythmic	Cardiovasuclar (sudden cardiac death)*
1992	Temafloxacin	Antibiotic	Blood and kidney
1997	Fenfuramine/Dexafluamine (used in combination since 1984; Fenfuramine Approved 1960)	Diet pill	Heart valve abnormalities
1998	Posico (Midefradil) (Approved 1996)	Ca⁺⁺ channel blocker	Lethal drug interactions (inhibited liver enzymes)
	Duract (Bronfemic Sodium) (early preapproval warnings of liver enzymes)	Pain relief	Liver damage
1999	Tronan (use severely restricted)	Antibiotic	Liver/kidney damage
	Raxar	Quinolone antibiotic	QT internal prolongation/ ventricular arrhythmias (deaths)*
	Hismanal	Antihistamine	Drug–drug interactions
	Rotashield	Rotavirus vaccine	Bowel obstruction
2000	Renzulin (approved Dec. 1996)	Type II diabetes	Liver damage
	Propulsid	Heartburn	Cardiovascular irregularities/ deaths*
	Lotonex	Irritable bowel syndrome	Ischemic colitis/ death*
2001	Phenylpropanolamine (PPA)	OTC ingredient	Hemorrhagic stroke
	Baychlor	Cholesterol reducing (satin)	Rhabdomyolysis (muscle-weakening) (deaths)

* Functional (pharmacologic) effects

Table 1.4 Classification of Adverse Drug Reactions (ADRs) in Humans

Type A	Dose-dependent; predictable from primary, secondary, and safety pharmacology	Main cause of ADRs (~75%), rarely lethal
Type B	Idiosyncratic response, not predictable, not dose-related	Responsible for ~25% of ADRs, but majority of lethal ones
Type C	Long-term adaptive changes	Commonly occurs with some classes of drug
Type D	Delayed effects, e.g., carcinogenicity, teratogencity	Low incidence
Type E	Rebound effects following discontinuation of therapy	Commonly occurs with some classes of drug

Note: Conventional safety pharmacology studies can only reasonably be expected to predict Type A ADRs. Functional toxicology measurements may predict Type C ADRs. Conventional toxicology studies address Type D ADRs. Prediction of Type B responses requires a more extensive preclinical and clinical evaluation, often only addressing risk factors for the idiosyncratic reponses (e.g., QT prolongation for torsades de pointes). Type E ADRs are rarely investigated preclinically using functional measurements unless there is cause for concern [34].

Adverse effects can be considered as being of one of five types (Table 1.4), of which only one is likely to be detected by conventional safety pharmacology testing.

It is recognized to be important that the tests employed detect bi-directional drug effects and that the tests performed be validated in both directions with appropriated reference (control) substances. This requirement is less appropriate for multiparameter procedures. When evaluations are subjective (as in most central nervous system, or CNS, evaluations), blinded testing should be considered. Ethical considerations (such as number of animals utilized) are important, but the ultimate ethical criterion is the assessment of risk for humans. Safety pharmacology studies should not be overinclusive but should be performed to the most exacting standards, including Good Laboratory Practice (GLP) compliance [35]. General pharmacology screens (see Table 1.5) are generally not conducted in accordance with GLPs, and thus the documentation, quality, and reproductivity can be suspect [36]. Tier I safety pharmacology data are acknowledged as essential to be available during the planning stage for Phase I studies, but this is only recently recognized to be the case and the data are not universally provided. This arises partly from the viewpoint that human tolerance (particularly in a well-designed and executed Phase I study in normal volunteers) is, in itself, an adequate assessment of general and safety pharmacology. This is, of course, backward—such human tolerance is rather, properly, an extension (and expression) of the nonclinical safety pharmacology.

Table 1.5 Differences between Safety Pharmacology and General Pharmacology

	Safety Pharm	General Pharm
Dose Range	Efficacious to Effect Level or Dose Producing Toxicity	Efficacious to Multiple of Efficacious Dose
Organizational Location	Development	Discovery
GLP	Yes	No
Objectives		
Identify Ancillary Actions / Side Effects	Yes	Yes
Establish Effect / No Effect Levels	Yes	No

1.3   Reasons for poor predictive performance

The other point of view in the past has been that properly executed repeated dose preclinical safety studies meeting the current design will (or could) fill these needs. Undesired pharmacological activities of novel drugs or biologicals may limit development of a therapeutic agent prior to the characterization of any toxicological effects. In rodent species, general pharmacology assays have traditionally been used to screen new agents for pharmacological effects on the central and peripheral nervous systems, the autonomic nervous system and smooth muscles, the respiratory and cardiovascular systems, the digestive system, and the physiological mechanisms of water and electrolyte balance. In large animal species, such as dogs and nonhuman primates, smaller numbers of animals per study limit their use for screening assays, but these species may play an important role in more detailed mechanistic studies. For drugs and biologicals that must be tested in nonhuman primates because of species-specific action of the test agent, functional pharmacology data are often collected during acute or subacute toxicity studies. This requires careful experimental design to minimize any impact that pharmacological effects or instrumentation may have on the assessment of toxicity. In addition, with many new therapies targeted at immunological diseases, the pharmacological effect of therapeutics on the immune system presents new challenges for pharmacology profiling. The applications of pharmacology assays by organ system in both rodent and large animal species are discussed, as well as practical issues in assessing pharmacological endpoints in the context of toxicity studies [37,38].

Various reasons why preclinical safety pharmacology tests may not accurately predict human adverse effects include the following:

- species differences in the presence or functionality of the molecular target mediating the adverse effect
- differences in absorption, distribution, metabolism, and excretion (ADME) between test species and man
- sensitivity of the test system—observations of a qualitative nature should be followed up with specific, quantitative assessment
- poor optimization of test conditions—the baseline level has to be set correctly to detect drug-induced changes
- study designs that are underpowered statistically
- inappropriate timing of functional measurements in relation to T_{max}
- delayed effects—safety pharmacology studies generally involve a single administration with time points for up to 24 h post-dose
- difficulty of detection in animals—adverse effects such as headache, disorientation, and hallucinations are quite a challenge to safety pharmacologists

In Europe, the numbers of registered drugs and the quantity of expenditure for drugs are increasing rapidly. Within the EU there are no longer any regulations requiring that new drugs have to be better than old ones. At the same time, pharmacoepidemiology studies in Europe and the United States show that adverse drug reactions now may account for up to 10% of the admissions of patients to hospitals at a cost of hundreds of millions of U.S. dollars annually [39]. This represents a considerable increase compared with the 1970s and before. A partial explanation is the many shortcomings of clinical trials and their relevance for health care. Adverse drug reactions are often poorly studied and documented in these studies and very seldom included in health economical analyses of the value of new drugs. Pharamcovigilance is product oriented—rather than utilization oriented— and quite invisible in clinical medicine. This is regrettable, since up to 50% of adverse drug reactions (ADRs) are dose dependent and thus preventable. Hopefully, the rapid progress in molecular and clinical pharmacogenetics will provide new tools for clinicians to choose and dose drugs according to the individual needs of patients. A good starting point for those not well versed in pharmacology and the range of potential mechanisms of action and of interaction can be found in Goodman and Gilman [40].

1.4   Why tiers?

One aspect of the International Conference on Harmonisation of Technical Requirements for Registration of Pharmaceuticals for Human Use (ICH) guidelines that confuses some people is the tier system, or the division of organ systems into sequential groups to be evaluated. Core Tier I, which must be performed prior to first studies in man, includes evaluations of central nervous system, cardiovascular system, and respiratory/

pulmonary system. Optional Tier II, which must be performed eventually prior to drug registration but be considered on a case-by-case basis prior to human trials, includes the other organ systems.

The rationale for this tier approach has to do with the judged relative hazard presented by effects on the different organ systems. Adverse function effects on core organ systems can be acutely fatal. Adverse effects on those organ systems that are optional, while of concern and potentially fatal, are generally unlikely to be acutely fatal and therefore are considered to not present a meaningful risk first in human trials. The exception to this is the immune system, but it is felt that the existent separate guideline for immunotoxicity evaluation calls for adequate steps to guard against acutely lethal effects on this system.

1.5   Study designs and principles

As a starting place, unlike older pharmacology studies, safety pharmacology studies are normally conducted as GLP studies. At the same time, unlike other safety assessment studies, these do not need to vastly exceed intended therapeutic doses so as to identify signs of toxicity. In this sense, they are closer to hazard tests. General considerations in the selection and design of safety pharmacology studies are straightforward as noted in this section.

1.5.1   Selection of methodology and species

The following selection criteria are worth considering for safety pharmacology studies:

1. It is preferable to use the same species for *in vivo* tests as those used in dystrophia myotonica protein kinase (DMPK) and toxicology—generally rat and dog (an exception would be when the primary pharmacological target in those species is different from the human form).
2. The methods should be well established (i.e., not a test invented internally that has not been subject to external evaluation).
3. The method/test should be in common use in research, and not one that is not broadly in current use.
4. The method/test should be validated in-house with at least one reference substance with known effect in humans.
5. They should give reliable, reproducible results every time.
6. The level of technical difficulty should be compatible with routine use in a pharmaceutical/CRO environment.*

---

* See Appendix A for list of acronyms.

General guidance for dose (or concentration) selection for such studies is as follows:

- *In vivo* studies:
  - Designed to define the dose response curve of the adverse effects.
  - Doses should include and exceed primary pharmacodynamic or therapeutic range.
  - In absence of safety pharmacology parameters, the highest doses equal or exceed some adverse effects (toxic range).
- *In vitro* studies:
- Generally designed to establish an effect-concentration relationship (range of concentrations).

Consideration in the selection and design of specific studies is straightforward.

- The following factors should be considered (selection):
  - Effects related to the therapeutic class
  - Adverse effects associated with members of the chemical/therapeutic class
  - Ligand binding or enzyme data suggesting a potential for adverse effects
  - Data from investigations that warrant further investigation
- A hierarchy of organ systems can be developed:
  - Importance with respect to life-supporting functions:
    - Cardiovascular
    - Respiratory
    - Central nervous system
  - Functions that can be transiently disrupted without causing irreversible harm:
    - Renal/urinary system
    - Autonomic nervous system
    - Gastrointestinal system
    - Other organ systems
  - No testing is usually considered necessary for the following:
    - Locally applied agents (e.g., dermal or ocular) where systemic exposure or distribution to the vital organs is low
    - Cytotoxic agents for treatment of end-stage cancer patients
    - Biotechnology-derived products that achieve highly specific receptor targeting (refer to toxicology studies)
- New salts having similar pharmacokinetics and pharmacodynamics

Table 1.6 presents a summary of guidance for the key design features for safety pharmacology studies.

Table 1.6 Summary of Key Study Design Features of the Safety Pharmacology Core Evaluation

Attribute	Recommendation	Comment
Animal model	Conscious, unrestrained, telemeterized, and/or trained	Should approximate conditions of Phase I clinical model
Test species	Rodent and/or nonrodent species used in toxicology studies or most appropriate species based upon scientific considerations	Generally, the toxicology species are the most appropriate, and specific synergies are available when these species are used.
Statistical design	Random blocked study (each animal receives all treatments and serves as own control)	Blocked study designs reduce animal used by up to 75% without loss of statistical power.
Group size	Sufficient to rule out a significant effect of the drug	Based upon laboratory experience with the model and a statistical power analysis
Controls	1. Negative (vehicle or placebo) 2. Positive	1. Always 2. Only when deemed necessary to ensure appropriate functionality of the test system
Route of administration	Best approximation of the clinical route	Should approximate conditions of Phase I clinical model
Test article formulation	Best approximation of the clinical formulation, given constraints posed by specific species	Formulations should be that used in toxicology and/or DMPK studies, so that bioavailability data may be used to assist in safety pharmacology study design and data interpretation.
Dose range	Three doses including a maximum tolerated dose	Dose-response requirements (S7A)
High dose	1. Maximum tolerated dose 2. Maximum feasible dose 3. Limit dose	1. Overlap toxicological dose range 2. Limited by formulation or dose volume constraints 3. 1–2 g/kg
Pharmacokinetics	Systemic exposure should be documented within the study or referenced to another relevant study	If SP study design is coordinated with toxicology program, toxicokinetic data may be used to establish safety pharmacology exposures.

Table 1.6 (Continued) Summary of Key Study Design Features of the Safety
Pharmacology Core Evaluation (from ICH S7A)

Attribute	Recommendation	Comment
Endpoints measured	Arterial blood pressure, heart rate, electrocardiogram	S7A requirements for cardiovascular core study
Duration	Generally performed as a single-dose study Sufficient to cover: 1. Initial distribution 2. Cmax 3. AUC phases (generally 24h)	1. Acute, transient (anaphylactoid) effects 2. Acute concentration-dependent effects 3. Delayed effects
Timing with respect to clinical development	Prior to first administration in humans	ICH S6: S7A Guidelines
Applicability of GLPs	Core battery studies should be performed to GLP standards to the greatest extent possible	Aspects of the study not conducted to GLPs should be identified and explained in the study protocol.

Source: ICH, *Safety Pharmacology Studies for Human Pharmaceuticals,* 2002.

1.6   Issues

The absence of observed activity may represent either a true or false negative effect. If an assay is valid for the particular test article and fails to indicate activity, it is an appropriate indicator of future events [41]. However, if the assay is insensitive or incapable of response, the test represents a form of bias, albeit unconscious. Many biological products demonstrate a specificity of response that limits the utility of commonly employed safety studies. Specificity for many biologies arises from both their physiochemical properties and their similarity to endogenous substances that are regulated in a carefully controlled manner. To overcome the issue of a lack of predictive value, various approaches may be used. For example, a multiple testing strategy of mutually reinforcing studies may be employed or safety studies may be adaptively fit to the biological circumstance.

A separate issue is how and when to consider isomers, metabolites, and the actual finished product:

- Generally parent compound and its major metabolite(s) that achieve systemic exposure should be evaluated.
- It may be important to test active metabolites from humans.
- Testing of individual isomers should also be considered.
- Studies with the finished product are only necessary if kinetics/dynamics are substantially altered in comparison to the active substance previously tested.

Special considerations are taken for statistically evaluating specific aspects of these studies. Specifically, analysis of time to event becomes very important [42].

1.7   Integral versus separate

There are two general approaches to the *in vivo* preclinical evaluation of safety pharmacology. Such evaluation can either be performed as (separate) free-standing studies focused solely on the pharmacologic endpoints of concern or as integral evaluations conducted as on animals that are part of a modified safety assessment study [38,43].

The argument for conducting free-standing studies is that (1) toxicology studies are generally conducted at a higher dose level than is the case of safety pharmacology studies, (2) special manipulations (such as implanted sensors) potentially desirable for safety pharmacology evaluations may not be practical in toxicity studies, and (3) the usually practiced daily dosing regimen can easily obscure pharmacologic effects and their relationships to systemic exposure.

These issues are subject to considerable disagreement. Certainly argument (1) is weak at best—while traditional toxicology studies do not include a group dosed at the projected clinical dose, the lowest dose is usually at a modest multiple of such a dose (5 or 10 times) within the range of doses desirable to be covered in a safety pharmacology study.

The second argument presupposes two points: (a) that special manipulations are required to adequately evaluate potential hazard, and (b) that required manipulations compromise the inclusion of involved animals in the regular toxicology evaluations. Point (a) seems a weak point at best, and likely not valid at all. Certainly in the case of the pre-IND central nervous system safety pharmacology evaluation, the proposed free-standing Irwin screen evaluation was the basis for the functional observational battery (FOB), which is standardly inducted in the rodent GLP toxicity studies and was developed for just this purpose [44]. Similarly, the EKG evaluation performed in the standard pivotal nonrodent (particularly dog) toxicology study that supports an IND should serve to provide a sensitive (certainly adequate) detector of any effects on heart rate or cardiac electrophysiological events. Only the regulatorily required evaluation of respiratory effects appears to not currently be addressed.

Point (b) is a stronger argument, though again not in the case of the CNS pharmacology evaluation. As Table 1.7 summarizes, the coverage of desired endpoints in an evaluation of cardiovascular and respiratory endpoints in current "standard" toxicology designs such shortcomings might be addressed with alterations in current study designs, but are not currently so. Finally, argument (3) is largely a matter of operational/logistical difficulty and potential difficulties in interpretation of results.

Table 1.7 Integration of Parameters into Safety Studies

Parameter	Safety pharmacology	Toxicology Rodent	Toxicology Nonrodent
Functional observational battery	++	++	+
Cardiovascular System	++	−	++
ECG			
Blood pressure	+	−	+
Systolic	+	+	−
Diastolic	+	−	−
LVP	+	−	−
Respiratory System	+++	++	—
Inspiration time	+	+	
Expiration time	+	plethysmography	
Peak inspiratory flow	+	in conscious	
Peak expiratory flow		unrestrained	
Respiratory rate	+	animals	
Tidal volume	+	+	
Resistance	+	−	
Compliance	+	−	
Hyperreactivity	+	−	

Note: +++, already integrated; ++, integration after adaptation easy; +, integration after adaptation possible; −, integration not feasible; ECG, electrocardiogram; LVP, lateral ventricular pressure.

The arguments for integration into existing study designs may also be viewed as a mixed bag. These are generally seen as (1) the desirability of avoiding the use of additional animals, the cost of such additional studies, and the need for additional drugs at an early stage of development; (2) the potential greater value and ease of interpretation of results when (potentially) isolated effects with one endpoint can be viewed in the context of the more thorough array of information captured in well-designed and well-executed GLP toxicity studies; and (3) the usually practiced daily dosing regimen can easily obscure pharmacologic effects and their relationship to systemic exposure.

Argument (1) against unwarranted use of animals is powerful and must be addressed. It should be remembered, however, that most safety pharmacology studies are not terminal—the animals employed are most commonly "washed out" and then reused in subsequent evaluations. The cost issue, while important, is less compelling when weighed against concerns of potential safety.

The integrated data evaluation argument (2) must also be carefully considered. It is, in a sense, the other side of the argument, which says that separate studies can best address the full range of functional endpoints of concern (as laid out in Table 1.7). Is one side a necessary and essential trade-off for the other?

Argument (3) comes from the practical logistics of dosing a large number of animals at one time, which makes careful observation/examination of each animal immediately afterward difficult (and even uncertain). It can lead to missed or erroneous observations. Currently the integration of FOB (CNS) evaluations into regular early toxicity studies is widely practiced, but not evaluation of the cardiovascular or respiratory endpoints.

1.8   Summary

The regulatorily mandated safety pharmacology (functional safety) studies as part of the drug development process are overdue and clearly serve to decrease potential undue hazards previously identified only postmarket in potentially large populations. However, the actual implementation of requirements and the use of the resulting data in risk/benefit decision continue to evolve.

References

1. Zbinden, G., *Pharmacological Methods in Toxicology*, Elmsford, NY: Pergamon, 1979, p. 613.
2. Kolata, G.B., The death of a research subject, *The Hastings Center Report*, 1980, 10:5–6.
3. Marshall, E., Volunteer's death prompts review, *Science*, 2001, 292:2226–2227.
4. Orme, M., et al., Healthy volunteer studies in Great Britain: the results of a survey into 12 months activity in this field, *By J. Clin. Pharm.*, 1989, 27:125–133.
5. Sibille, M., et al., Adverse events in phase-I studies: a report in 1015 healthy volunteers, *Eur. J. Clin. Pharmacol.*, 1998, 54:13–20.
6. Darragh, A., et al., Sudden death of a volunteer, *Lancet*, 1985, 1:93–94.
7. Irwin, S., Drug screening and evaluation of new compounds in animals. In: *Animal and Clinical Pharmacologic Techniques in Drug Evaluations*, (Nodine, H., and Siegler, P.E., eds.). Philadelphia: Year Book Medical, 1964, pp. 64–76.
8. Turner, R.A., *Screening Methods in Pharmacology*, vols. I and II, New York: Academic, 1965, pp. 42–47, 60–68, 27–128.
9. Bramm, E., Binderup, L., and Arrigoni-Martelli, E., An unusual profile of activity of a new basic anti-inflammatory drug, timegadine, *Agents Actions*, 1981, 11:402–409.
10. Graf, E., et al., Animal experiments on the safety pharmacology of lofexidine, *Arzneim.-Forsch./Drug Res.*, 1982, 32(II)(8a):931–940.
11. Lumley, C.E., General pharmacology, the international regulatory environment, and harmonization of guidelines. *Drug Dev. Res.*, 1994, 32:223–232.

12. Proakis, A.G., Regulatory considerations on the role of general pharmacology studies in the development of therapeutic agents. *Drug Dev Res.*, 1994, 32:233–36.

13. Igarashi, T., Nakane, S., and Kitagawa, T., Predictability of clinical adverse reactions of drugs by general pharmacology studies, *J. Toxicol. Sci.*, 1995, 20:77–92.

14. Hite, M., Safety pharmacology approaches, *Int. J. Toxicol.*, 1995, 16:23–31.

15. Folke, S., Drug safety in relation to efficacy: the view of a clinical pharmacologist, *Pharmacol. Toxicol.*, 2000, 86:30–32.

16. Fujimori, K., The role of general pharmacological studies and pharmacokinetics in the evolution of drugs (1): The role of general/safety pharmacology studies in the development of pharmaceuticals: International harmonization guidelines, *Folia Pharmacologica Japonica.*, 1999, 13:31–39.

17. Thompson, E.B., *Drug Bioscreening: Drug Evaluation Techniques in Pharmacology*, New York: VCH Publishers, Inc., 1990, p. 366.

18. Sills, M., *In vitro* screens and functional assays to assess receptor pharmacology, *Drug Dev. Res.*, 1994, 32:260–268.

19. Zbinden, G., The significance of pharmacologic screening tests in the preclinical safety evaluation of new drugs, *J. New Drugs*, 1966, 6:1–7.

20. Zbinden, G., Neglect of function and obsession with structure in toxicity testing. In: *Proc. 9th Int. Cong. Pharmacol.*, vol. 1, New York: Macmillan, 1984, pp. 43–49.

21. Williams, P.D., The role of pharmacological profiling in safety assessment, *Reg. Toxicol. Pharmacol.*, 1990, 12:238–252.

22. Williams, P.D., Proposal for a core battery-package for phase I. In: Workshop on *In the Use of Pharmacology Studies in Drug Safety Assessment: Present Situation and Future Perspectives* (Sundwall, A., et al., eds.), Tryckgruppen, Stockholm, Sept. 26–27, 1994, pp. 133–138.

23. Anon, *Guidelines for the safety pharmacology study required for application for approval of manufacturing (import) of new drugs: Notification No. 4 of 29th January.* Director of New Drug Division, Pharmaceutical Affairs Bureau, Japan Ministry of Health and Welfare, Takushin-Yaku, 1991.

24. Anon, Guidelines for general pharmacology. In: *Drug Approval and Licensing Procedures in Japan 1992*, Tokyo: Yakugyo Jiho, 1992, pp. 137–140.

25. Kinter, L.B., Gossett, K.A., and Kerns, W.D., Status of safety pharmacology in the pharmaceutical industry, 1993, *Drug Dev. Res.*, 1994, 32:208–216.

26. Sullivan, A.T., and Kinter, L.B., Status of safety pharmacology in the pharmaceutical industry—1995, *Drug Dev. Res.*, 1995, 12:238–252.

27. Kurata, M., Kanai, K., and Mizuguchi, K., Trends in safety pharmacology in the U.S. and Europe, *J. Toxicol. Sci.*, 1997, 22:237–248.

28. Olejiniczak, K., Development of a safety pharmacology guideline, *Human Exper. Toxicol.*, 1999, 18:502.

29. CPMP, *Note for Guidance on Safety Pharmacology Studies in Medicinal Product Development*, 1998.

30. ICH, *ICH S7A Safety Pharmacology Studies for Human Pharmaceuticals*, 2002.

31. ICH, *Safety Pharmacology Studies for Assessing the Potential for Delayed Ventricular Repolarization (QT Interval Prolongation) by Human Pharmaceuticals*, 2000.

32. Claude, J.R., Safety pharmacology in the nonclinical assessment of new medicinal products: definition, place, interest, and difficulties, *Fund. Clin. Pharmacol.*, 2002, 16:75–78.

33. Gad, S.C., *Drug Safety Evaluation,* 2nd edition, New York: John Wiley & Sons, 2009.
34. Redfern, W.S., et al., Safety pharmacology: a progressive approach, *Fund. Clin. Pharmacol.,* 2002, 16:161–173.
35. Anon, *Applicability of Good Laboratory Practices,* Committee on Proprietary Medicinal Products III (3824/92): Rev. 1, item 9, 1992.
36. Spindler, P., and Seiler, J.P., The quality management of pharmacology and safety pharmacology studies, *Fund. Clin. Pharmacol.,* 2002, 16:83–90.
37. Martin, L.I., Horvath, C.J., and Wyand, M.S., Safety pharmacology screening: practical problems in drug development, *Int. J. Toxicol.,* 1997, 16:41–65.
38. Matsuzawa, T., et al., Current status of conducting function tests in repeated dose toxicity studies in Japan, *J. Toxicol. Sci.,* 1997, 22:374–382.
39. Sjouist, F., Drug-related hospital admissions, *Ann. Pharmacol.,* 2000, 34:832–839.
40. Brunton, L.L., Lazo, J.S., and Parker K.L., *Goodman & Gilman's The Pharmacological Basis of Therapeutics,* 11th ed., New York: McGraw-Hill, 2006.
41. Green, M.D., Problems associated with the absence of activity in standard models of safety pharmacology used to assess biological products, *Int. J. Toxicol.,* 1997, 16:33–40.
42. Anderson, H., et al., Statistical analysis of time to event data from preclinical safety pharmacology studies, *Tox. Methods,* 2000, 10:111–125.
43. Luft, J., and Bode, G., Integration of safety pharmacology endpoints into toxicology studies, *Fund. Clin. Pharmacol.,* 2002, 16:91–103.
44. Gad, S.C., A neuromuscular screen for use in industrial toxicology, *J. Toxicol. Environ. Health,* 1982, 9:691–704.

chapter two

Regulatory requirements
ICH, FDA, EMA, and Japan

2.1   Regulatory requirements

The International Conference on Harmonisation of Technical Requirements for Registration of Pharmaceuticals for Human Use (ICH S7A) guidelines requiring safety pharmacology evaluations of new pharmaceuticals came into force in the middle of 2001, preceded by various national guidelines and followed by other ICH guidances (see Table 2.1). These regulations have evolved and been modified, but there are still regional variations as to interpretation of the requirements.

Japan operated in conformance to the Ministry of Health and Welfare (MHW) draft (*Guidelines for Safety Pharmacology Studies*, revised in 1999 [1]) prior to full ICH implementation. The basic principle of the revision is to harmonize the guideline with the international concepts. The working group decided to change the wording in the title from *General Pharmacology* to *Safety Pharmacology* because the objective of this guideline is to assess the safety of a test substance in humans by examining the pharmacodynamic properties of the substance (note that Japan has separate guidelines for general pharmacology [2]). The proposed guideline includes studies on vital functions as essential studies that should be performed prior to human exposure. Studies are also required to be conducted when predictable or unexpected observed effects are concerned. The working group recommended a case-by-case approach to select the necessary test items in consideration of the variable information available, which is not the primary ICH approach.

In the European Union (EU), the Committee on Proprietary Medicinal Products (CPMP) issued a draft *Note for Guidance on Safety Pharmacology Studies in Medicinal Product Development* in 1997 [3], but it was not finalized or put in force until the middle of 2001. The U.S. Food and Drug Administration (FDA) promulgated equivalent guidance at the same time, but the exact details of compliance and implementation, as will be seen in this volume, are still being worked out.

A further consideration arose in 2008 with the promulgation of guidance requiring safety assessment of significant human metabolites including their safety pharmacology aspects [4].

Table 2.1 Regulatory Guidelines for Safety Pharmacology

	ICH	FDA	EMA/CPMP	JAPAN/MHW
Safety assessment of pharmaceuticals	M3 Nonclinical Safety Studies for the Conduct of Human Clinical Trials for Pharmaceuticals [6]	Guidance for Industry: Nonclinical Safety Studies for the Conduct of Human Clinical Trials for Pharmaceuticals [7]	N/A	New Drugs Division Notification No. 9/99: Guidelines for Toxicity Studies of Drugs
Safety assessment of biotech therapeutics	S6 Preclinical Safety Evaluation of Biotechnology-Derived Pharmaceuticals [8]	Guidance for Industry: Providing Clinical Evidence of Effectiveness for Human Drug and Biological Products	Note for Guidance on Comparability of Medicinal Products Containing Biotechnology-Derived Proteins as a Drug Substance	N/A
Safety pharmacology	Guidance for Industry: S7A Safety Pharmacology Studies for Human Pharmaceuticals [5]	Guidance for Industry: S7A Safety Pharmacology Studies for Human Pharmaceuticals [4]	CPMP: Not for Guidance on Safety Pharmacology Studies in Medicinal Product Development [3]	Notification No. 4: Guidelines for General Pharmacology [1]
QT interval	Safety Pharmacology Studies for Assessing the Potential for Delayed Ventricular Repolarization (QT Interval Prolongation) by Human Pharmaceuticals [9]	N/A	Points to Consider: The Assessment of the Potential for QT Interval Prolongation by Noncardiovascular Medicinal Products	N/A

Source: FDA, *Guidance for Industry: Nonclinical Studies for Development of Pharmaceutical Excipients.* http://www.fda.gov/cder/guidance/3812dft.pdf, 2002.

Table 2.2 Cardiovascular System Safety Pharmacology Evaluations

Core

- Hemodynamics (blood pressure, heart rate)
- Autonomic function (cardiovascular challenge)
- Electrophysiology (EKG in dog)

QT prolongation (noncore)

An additional guideline, ICH S7B, has been promulgated and addresses the assessment of potential for QT prolongation. In the meantime, CPMP 986/96 indicates the following preclinical studies should be conducted prior to first administration to man:

- Cardiac action potential *in vitro*
- ECG (QT measurements) in a cardiovascular study which would be covered in the core battery
- HERG channel interactions (HERG expressed in HEK 293 cells)

Source: ICH, *ICH S7B Safety Pharmacology Studies for Assessing the Potential for Delayed Ventricular Repolarization (QT Internal Prolongation) by Human Pharmaceuticals.* http://www.ich.org/pdfICH/S7B.pdf, 2004

Table 2.3 Respiratory System Safety Pharmacology Evaluation

Respiratory functions

Measurement of rate and relative tidal volume in conscious animals

Pulmonary function

Measurement of rate, tidal volume, and lung resistance and compliance in anesthetized animals

The actual requirements of the final June 2001 ICH guidelines [5] are broadly sketched. They call for the conduct of studies in a core battery to assess effects on the cardiovascular (Table 2.2), respiratory (Table 2.3), central nervous (Table 2.4), and secondary organ (Table 2.5) systems. Follow-up studies for the core battery are also required on a case-by-case basis for the three main organ systems, the functional compromise of which is considered immediately life threatening:

- Central Nervous System (CNS)
  - Behavioral pharmacology, learning and memory, specific ligand binding, neurochemistry, visual, auditory and/or electrophysiology examinations
- Cardiovascular System (CV)
  - Hemodynamics (blood pressure and heart rate), electrophysiology (EKG in dog and other such screens for QTc prolongation as appropriate), and autonomic function in response to a pharmacologic challenge
  - An *in vitro* assay (the hERG) is also recommended

Table 2.4 Central Nervous System (CNS) Safety Pharmacology Evaluation

---

Irwin test: General assessment of effects on gross behavior and physiological state*

Locomotor activity: Specific test for sedative, excitatory effects of compounds

Neuromuscular function: Assessment of grip strength

Rotarod: Test of motor coordination

Anesthetic interactions: Test for central interaction with barbiturates

Anti-/proconvulsant activity: Potentiation or inhibition of effects of pentylenetetrazole

Tail flick: Tests for modulation of nociception (also hot plate, Randall Selitto, tail pinch)

Body temperature: Measurement of effects on thermoregulation

Autonomic function: Interaction with autonomic neurotransmitters *in vitro* or *in vivo*

Drug dependency: Test for physical dependence, tolerance, and substitution potential

Learning and memory: Measurement of learning ability and cognitive function in rats

---

Note: Tier II or secondary organ system requirements are less precisely presented, as Table 2.5 makes clear.

* Usually a functional observational battery (FOB) is integrated into a rodent (rat); repeat dose toxicity studies to meet this requirement.

Table 2.5 Secondary Organ System Safety Pharmacology Evaluation

---

Renal system

Renal Function—Measurement of effects on urine excretion in saline-loaded rats

Renal Dynamics—Measurement of renal blood flow, glomerular filtration rate, and clearance

Gastrointestinal (GI) system

GI Function—Measurement of gastric emptying and intestinal transit

Acid Secretion—Measurement of gastric acid secretion (Shay rat)

GI Irritation—Assessment of potential irritancy to the gastric mucosa

Emesis—Nausea, vomiting

Immune system

Passive Cutaneous Anaphylaxis (PCA)—Test for potential antigenicity of compounds

Other

Blood Coagulation

In Vitro Platelet Aggregation

In Vitro Hemolysis

- Respiratory System (Respiratory/Pulmonary)
- Tidal volume, bronchial resistance, compliance, pulmonary arterial pressure, blood gases

There are conditions defined under which safety pharmacology studies are not necessary:

- Locally applied agents (e.g., dermal or ocular) where systemic exposure or distribution to the vital organs is low
- Cytotoxic agents for treatment of end-stage cancer patients
- Biotechnology-derived products that achieve highly specific receptor targeting (refer to toxicology studies)
- New salts having similar pharmacokinetics and pharmacodynamics

The FDA has, meanwhile, published draft guidance [10] calling for safety pharmacology evaluation of all new pharmaceutical excipients. Additionally, previously approved agents (that did not have safety pharmacology assessments due to predating the requirements) for which a new use or formulation application (via the 505(b)(2) route) is being made are generally being required to meet current safety pharmacology standards.

References

1. Anon, *Guidelines for the Safety Pharmacology Study Required for Application for Approval of Manufacturing (Import) of New Drugs: Notification No. 4.*, Japan Ministry of Health and Welfare, January 29, 1991.
2. Anon, Guidelines for general pharmacology. In: *Drug Approval and Licensing Procedures in Japan, 1992*, Japanese Ministry of Health and Welfare, pp. 137–140.
3. CPMP, *Note for Guidance on Safety Pharmacology Studies in Medicinal Product Development*, 1998.
4. CDER, *Guidance for Safety Testing of Drug Metabolites*, 2008.
5. FDA, *ICH S7A Guidance for Industry: Safety Pharmacology Studies for Human Pharmaceuticals.* http://www.fda.gov/cder/guidance/4461fnl.pdf, 2001.
6. ICH, *ICH S7A (Step 4) ICH Harmonized Tripartite Guidelines on Safety Pharmacology Studies for Human Pharmaceuticals.* http://www.ich.org/pdfICH/ S7step4.pdf, 2000.
7. ICH, *ICH Harmonized Tripartite Guidelines (M3): Timing of Non-clinical Safety Studies for the Conduct of Human Clinical Trials for Pharmaceuticals.* 2009.
8. ICH, *M3 (R2) Nonclinical Safety Studies for the Conduct of Human Clinical Trials for Pharmaceuticals.* http://www.fda.gov/cder/guidance/1855fnl.pdf, 2009.
9. ICH, *E14: Clinical Evaluation of QT/QTc Interval Prolongation and Proarrythmic Potential for Non-Antiarrhythmic Drugs.* http://private.ich.org/LOB/media/MEDIA1476.pdf, 2005.
10. ICH, *ICH Harmonized Tripartite Guidelines (S6): Preclinical Safety Evaluation of Biotechnology-derived Pharmaceuticals.* http://www.fda.gov/cder/guidance/1859fnl.pdf, 1997b.

11. ICH, *ICH S7B Safety Pharmacology Studies for Assessing the Potential for Delayed Ventricular Repolarization (QT Internal Prolongation) by Human Pharmaceuticals.* http://www.ich.org/pdfICH/S7B.pdf, 2004.
12. FDA, *Guidance for Industry: Nonclinical Studies for Development of Pharmaceutical Excipients.* http://www.fda.gov/cder/guidance/3812dft.pdf, 2002.

Web Sources: FDA: www.fda.gov
                    CPMP: www.eudra.org
                    MHW: www.mhw.go.jp/english
                    ICH: www.ifpma.org

chapter three

Principles of screening and study design

3.1 Introduction

Screens are tests or searches performed with high sensitivity (but limited specificity) to detect the presence (or ensure the absence) of something. They are not generally definitive in the sense of characterizing what is found. In biological research, screens are tests designed and performed to identify agents or organisms having a certain set of characteristics that will either exclude them from further consideration or cause them to be selected for closer attention. In pharmaceutical safety assessment (and safety pharmacology, in particular), our use of screens is usually negative (i.e., no activity is found)—agents or objects possessing other-than-desired pharmacological or biochemical activities are considered to present enough of a hazard due to these secondary pharmacologic effects that they are not studied further or developed as potential therapeutic agents without compelling reasons (e.g., cases of extreme benefit such as lifesaving qualities). For safety pharmacology in particular, it is screens rather than definitive studies that are required and intended upon.

In the broadest terms, all of what is done in preclinical (and, indeed, in Phase I clinical) studies can be considered a form of screening [1]. This is certainly true of safety pharmacology studies. What varies is the degree of effectiveness of (or our confidence in) each of the tests used. As a general rule, even though we think of the expensive and labor-intensive "pivotal" studies required to support regulatory requirements (4-week to 1-year toxicity studies, carcinogenicity, and segment I–III studies, etc.) as definitive, in fact, they are generally effective but not necessarily efficient screens.

Though toxicologists and pharmacologists in the pharmaceutical industry are familiar with the broad concepts of screening, they generally do not recognize the applicability of screens. The principles underlying screening are also not generally well recognized or understood. The objective behind the entire safety assessment process in the pharmaceutical industry is to identify those compounds for which the risk of harming humans does not exceed the potential benefit to them. In most cases this means that if a test or screen identifies a level of risk that we have confidence in (our *activity criterion*), then the compound that was tested is no longer considered a viable candidate for development. In this approach, what may change from test to test is the activity criterion (i.e., our basis for and degree of confidence in the outcome). The goal of safety pharmacology is to minimize the number of false negatives in safety assessment. Anderson and Hauck [2] should be consulted for statistical methods to minimize false-negative results.

Figure 3.1 illustrates how, in the current climate, decisions are more likely to be made on a multidimensional basis, which creates a need for balance between (1) degree of benefit, (2) confidence that there *is* a benefit (efficacy is being evaluated in *models* or screens at the same time safety is), (3) type of risk (with, e.g., muscle irritation, mutagenicity, acute lethality,

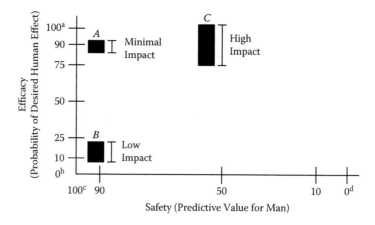

Figure 3.1 Decision making for pharmaceutical candidates based on outcome of screening tests. (a) A 100% probability of efficacy means that every compound that has the observed performance in the model(s) used has the desired activity in man. (b) A 0% probability of efficacy means that every compound that has the observed performance in the model(s) used does not have the desired activity in man. (c) A 100% probability of a safety finding means that such a compound would definitely cause this toxicity in man. (d) A 0% probability means this will never cause such a problem in man. Note: These four cases (a, b, c, and d) are almost never found. The height of the *impact* column refers to the relative importance (human risk) of a safety finding. Compound A has a high probability of efficacy but also a high probability of having some adverse effect in man. But if that adverse effect is of low impact—say, transitory muscle irritation for a life-saving antibiotic—then A should go forward. Likewise, B, which has low probability of efficacy and high probability of having an adverse effect with moderate impact, should not be pursued. Compound C is at a place where the high end of the impact scale should be considered. Though there is only a 50% probability of this finding (say, neurotoxicity or carcinogenicity) being predictive in man, the adverse effect is not an acceptable one. Here a more definitive test is called for or the compound should be dropped.

and carcinogenicity having various degrees of concern attached to them), and (4) confidence in and degree of risk. This necessity for balance is commonly missed by many who voice opposition to screens because they may cause us to throw out a promising compound based on a finding in which we have only (for example) 80% confidence. Screens, particularly those performed early in the research and development process, should be viewed as the biological equivalent of exploratory data analysis. They should be very sensitive, which by definition means that they will have a lot of *noise* associated with them. Screens generally do not establish that an agent is (or is not) a *bad actor* for a certain end point. Rather, they confirm that if interest in a compound is sufficient, a more definitive test (a confirmatory test) is required, which frequently will provide a basis for selecting between multiple candidate compounds.

3.2   Characteristics of screens

The terminology involved in screen design and evaluation and the characteristics of a screen should be clearly stated and understood. The characteristics of screen performance are defined as follows:

- Sensitivity: the ratio of true positives to total actives
- Specificity: the ratio of true negatives to total inactives
- Positive accuracy: the ratio of true to observed positives
- Negative accuracy: the ratio of true to observed negatives
- Capacity: the number of compounds that can be evaluated
- Reproducibility: the probability that a screen will produce the same results at another time (and, perhaps, in some other lab)

Later we will contrast some of these characteristics with key elements of study design.

These characteristics may be optimized for a particular use, if we also consider both the mathematics and the *errors* underlying them.

A brief review of the basic relationships between error types and statistical power starts with considering each of five interacting factors [3–5] that serve to determine power and define competing error rates.

α, the probability of our committing a type I error (a false positive)
β, the probability of our committing a type II error (a false negative)
Δ, the desired sensitivity in a screen (such as being able to detect an increase of 10% in mutations in a population)
σ, the variability of the biological system and the effects of chance errors
n, the necessary sample size needed to achieve the desired levels of each of these factors. We can, by our actions, generally change only this aspect (component) of the equation, since n is proportional to

$$\frac{\sigma}{\alpha, \ \beta, \ \text{and} \ \Delta}$$

The implications of this are, therefore, that (1) the greater σ is, the larger n must be to achieve the desired levels of α, β, and/or Δ; and (2) the smaller the desired levels of α, β, and/or Δ, if n is constant, the larger σ must be.

What are the background response level and the variability in our technique? As any good toxicologist will acknowledge, matched concurrent control (or standardization) groups are essential to minimize within-group variability as an error contributor. Unfortunately, in *in vivo* toxicology test systems, large sample sizes are not readily attainable, and there are other complications to this problem that we shall consider later.

In an early screen, a relatively large number of compounds will be tested. It is unlikely that one will stand out so much as to have greater statistical significance than all the other compounds [6]. Instead, a more or less continuous range of activities will be found. Compounds showing the highest (beneficial) or lowest (adverse) activity will proceed to the next assay or tier of tests in the series and may be used as lead compounds in a new cycle of testing and evaluation.

The balance between how well a screen discovers activities of interest versus other effects (specificity) is thus critical. Table 3.1 presents a graphic illustration of the dynamic relationship between discovery and discrimination.

Both discovery and discrimination in screens hinge on the decision criterion that is used to determine if activity has or has not been detected. How sharply such a criterion is defined and how well it reflects the working of a screening system are two of the critical factors driving screen design.

An advantage of testing many compounds is that it gives the opportunity to average activity evidence over structural classes or to study quantitative structure–activity relationships (QSARs), which can be used

Table 3.1 Discovery and Discrimination of Toxicants

Screen outcome	Actual activity of agent tested	
	Positive	Negative
Positive	a	b
Negative	c	d

Note: Discovery (sensitivity) = a/(a + c),
  where a = all toxicants found positive
    a + c = all toxicants tested.

  Discrimination (specificity) = d/(b + d),
  where d = all nontoxicants found negative
    b + d = all nontoxicants tested.

to predict the activity of new compounds and thus reduce the chance of *in vivo* testing on negative compounds. The use of QSARs can increase the proportion of truly active compounds passing through the system.

It should be remembered that maximization of the performance of a series of screening assays requires close collaboration among the toxicologist, chemist, and statistician. Screening, however, forms only part of a much larger research and development context. Screens thus may be considered the biological equivalent of exploratory data analysis (EDA). EDA methods, in fact, provide a number of useful possibilities for less rigid and yet utilitarian approaches to the statistical analysis of the data from screens and are one of the alternative approaches presented and evaluated here [7–10]. Over the years, the author has published and consulted on a large number of screening studies and projects. These have usually been directed at detecting or identifying potential behavioral toxicants or neurotoxicants, but some have been directed at pharmacological, immunotoxic, and genotoxic agents [11,12].

The general principles or considerations for screening in safety assessments are as follows:

1. Screens almost always focus on detecting a single point of effect (such as mutagenicity, lethality, neurotoxicity, or developmental toxicity) and have a particular set of operating characteristics in common.
2. A large number of compounds are evaluated, so ease and speed of performance (which may also be considered efficiency) are very desirable characteristics.
3. The screen must be very sensitive in its detection of potential effective agents. An absolute minimum of active agents should escape detection; that is, there should be very few false negatives (in other words, the type II error rate or beta (β) level should be low). Stated yet another way, the signal gain should be way up.
4. It is desirable that the number of false positives be small (i.e., there should be a low type I error rate or alpha (α) level).
5. Items 2–4, which are all to some degree contradictory, require the involved researchers to agree on a set of compromises, starting with the acceptance of a relatively high alpha level (0.10 or more), that is, a higher noise level.
6. In an effort to better serve item 1, safety assessment screens frequently are performed in batteries so that multiple end points are measured in the same operation. Additionally, such measurements may be repeated over a period of time in each model as a means of supporting item 2.
7. The screen should use small amounts of compound to make item 1 possible and should allow evaluation of materials that have limited availability (such as novel compounds) early on in development.

8. Any screening system should be validated initially using a set of blind (positive and negative) controls. These blind controls should also be evaluated in the screening system on a regular basis to ensure continuing proper operation of the screen. As such, the analysis techniques used here can then be used to ensure the quality or modify performance of a screening system.
9. The more that is known about the activity of interest, the more specific the form of screen that can be employed. As specificity increases, so should sensitivity. However, generally the size of what constitutes a meaningful change (that is, the Δ) must be estimated and is rarely truly known.
10. Sample (group) sizes are generally small.
11. The data tend to be imprecisely gathered (often because researchers are unsure what they are looking for) and therefore possess extreme within-group variability or modify test performance.
12. Proper dose selection is essential for effective and efficient screen design and conduct. If insufficient data are available, a suitably broad range of doses must be evaluated (however, this technique is undesirable on multiple grounds, as has already been pointed out).

Much of the mathematics involved in calculating screen characteristics came from World War II military-based operations analysis and research, where it was important for design of radar, antiair, and antisubmarine warfare systems and operations [13].

3.3   Uses of screens

The use of screens that first occurs to most pharmaceutical scientists is in pharmacology [14]. Early experiences with the biological effects of a new molecule are almost always in some form of efficacy or pharmacology screen. The earliest of these tend to be with narrowly focused models, not infrequently performed *in vitro*. The later pharmacology screens, performed *in vivo* to increase confidence in the therapeutic potential of a new agent or to characterize its other activities (e.g., cardiovascular, central nervous system), can frequently provide some information of use in safety assessment also (even if only to narrow the limits of doses to be evaluated), and the results of these screens should be considered in early planning. In the new millennium, requirements for specific safety pharmacology screens have been promulgated. Additionally, since the late 1990s two new areas of screening have become very important in pharmaceutical safety assessment. The first is the use of screens for detecting compounds with the potential to cause fatal cardiac arrhythmias. These are almost always preceded by the early induction of a prolongation of the QT interval. While this should be detected in the EKGs performed in repeat-dose canine studies, several early screens (such as the human *ether-a-go-go*-related gene, or hERG, screen) are more rapid and efficient (though not conclusive) for selecting candidate compounds for further development.

The other area of screen usage is with microassays in toxicogenomic screening—early detection of the potential for compounds to alter gene expressions with adverse consequences [15,16].

Safety assessment screens are performed in three major settings—discovery support, development (what is generally considered the "real job" of safety assessment), and occupational health / environmental assessment testing. Discovery support is the most natural area of employment of screens and is the place where effective and efficient screen design and conduct can pay the greatest long-range benefits. If compounds with unacceptable safety profiles can be identified before substantial resources are invested in them—and structures modified to maintain efficacy while avoiding early safety concerns—then long-term success of the entire research and development effort is enhanced. In the discovery support phase, one has the greatest flexibility in the design and use of screens. Here screens truly are used to select from among a number of compounds.

Examples of the use of screens in the development stage are presented in some detail in Section 3.7.1 on the central nervous system.

The use of screens in environmental assessment and occupational health is fairly straightforward. On the occupational side the concerns address the potential hazards to those involved in making the bulk drug. The need to address potential environmental concerns covers both true environmental items (aquatic toxicity, etc.) and potential health concerns for environmental exposures of drug into workers. The resulting work tends to be either regulatorily defined tests (for aquatic toxicity) or defined end points such as dermal irritation and sensitization, which have been (in a sense) screened for already in other nonspecific tests.

The most readily recognized examples of screens in toxicology are those that focus on a single end point. The traditional members of this group include genotoxicity tests; lethality tests (particularly recognizable as a screen when in the form of limit tests); and tests for corrosion, irritation (both eye and skin), and skin sensitization. Others that fit this same pattern, as will be shown, include the carcinogenicity bioassay (especially the transgenic mouse models) and developmental toxicity studies.

The *chronic* rodent carcinogenicity bioassay is thought of as the "gold standard" or definitive study for carcinogenicity, but, in fact, it was originally designed as (and functions as) a screen for strong carcinogens [17]. It uses high doses to increase its sensitivity in detecting an effect in a small sample of animals. The model system (be it rats or mice) has significant background problems of interpretation. As with most screens, the design has been optimized (by using inbred animals, high doses, etc.) to detect one type of toxicant—strong carcinogens. Indeed, a negative finding does not mean that a material is not a carcinogen but rather than it is unlikely to be a potent one.

Many of the studies done in safety assessment are multiple end point screens. Such study types as a 90-day toxicity study or immunotox/neurotox screens are designed to measure multiple end points with the desire of increasing both sensitivity and reliability (by correspondence/correlation checks between multiple data sets).

3.4   Types of screens

There are three major types of screen designs: the single stage, sequential, and tiered. Both the sequential and tiered are multistage approaches, and each of these types also varies in terms of how many parameters are measured. But these three major types can be characterized as follows:

Single Stage. A single test will be used to determine acceptance or rejection of a test material. After an activity criterion (such as X score in a righting reflex test) is established, compounds are evaluated based on being less than X (i.e., negative) or equal to or greater than X (i.e., positive). As more data are accumulated, the criterion should be reassessed.

Sequential. Two or more repetitions of the same test are performed, one after the other, with the severity of the criterion for activity being increased in each sequential stage. This procedure permits classification of compounds in to a set of various ranges of potencies. As a general rule, it appears that a two-stage procedure, by optimizing decision rules and rescreening compounds before declaring compounds *interesting*, increases both sensitivity and positive accuracy; however, efficiency is decreased (or its throughput rate).

Tiered (or Multistage). In this procedure, materials found active in a screen are reevaluated in one or more additional screens or tests that have greater discrimination. Each subsequent screen or test is both more definitive and almost always more expensive.

For purposes of our discussion here, we will primarily focus on the single-stage system, which is the simplest. The approaches presented here are appropriate for use in any of these screening systems, although establishment of activity criteria becomes more complicated in successive screens. Clearly, the use of multistage screens presents an opportunity to obtain increased benefits from the use of earlier (lower-order) screening data to modify subsequent screen performance and the activity criterion.

3.5   Criterion: Development and use

In any early screen, a relatively large number of compounds will be evaluated with the expectation that a minority will be active. It is unlikely that any one will stand out so much as to have greater statistical significance than all the other compounds based on a formal statistical test. A more or less continuous range of activities will be found. Compounds displaying a certain degree of activity will be identified as *active* and handled as such. For safety screens, those that are *inactive* go on to the next test in a series and may be used as lead compounds in a new cycle of testing and evaluation. The single most critical part of the use of screens is how to make the decision that activity has been found.

Each test or assay has an associated activity criterion. If the result for a particular test compound meets this criterion, the compound is active and handled accordingly. Such a criterion could have a statistical basis (e.g., all compounds with observed activities significantly greater than the control at the 5% level could be tagged). However, for early screens, a statistical criterion may be too strict, given the power of the assay, resulting in a few compounds being identified as active. In fact, a criterion should be established (and perhaps modified over time) to provide a desired degree of confidence in the predictive value of the screen.

A useful indicator of the efficiency of an assay series is the frequency of discovery of truly active compounds. This is related to the probability of discovery and to the degree of risk (hazard to health) associated with an active compound passing a screen undetected. These two factors, in turn, depend on the distribution of activities in the series of compounds being tested, and the chances of rejecting and accepting compounds with given activities at each stage.

Statistical modeling of the assay system may lead to the improvement of the design of the system by reducing the interval between discoveries of active compounds. The objectives behind a screen and considerations of (1) costs for producing compounds and testing, and (2) the degree of uncertainty about test performance will determine desired performance characteristics of specific cases. In the most common case of early toxicity screens performed to remove possible problem compounds, preliminary results suggest that it may be beneficial to increase the number of compounds tested, decrease the number of animals (or other test models) per assay, and increase the range and number of doses. The result will be less information on more structures, but there will be an overall increase in the frequency of discovery of active compounds (assuming that truly active compounds are entering the system at a random and steady rate).

The methods described here are well suited to analyzing screening data when the interest is truly in detecting the absence of an effect with little chance of false negatives. Many forms of graphical analysis methods are available, including some newer forms that are particularly well suited to multivariate data (the type that are common in more complicated screening test designs). It is intended that these aspects of analysis will be focused on in a later publication.

The design of each assay and the choice of the activity criterion should, therefore, be adjusted, bearing in mind the relative costs of retaining false positives and rejecting false negatives [18]. Decreasing the group sizes in the early assays reduced the chance of obtaining significance at any particular level (such as 5%), so that the activity criterion must be relaxed, in a statistical sense, to allow more compounds through. At some stage, however, it becomes too expensive to continue screening many false positives, and the criterion must be tightened accordingly. Where the criterion is set depends on what acceptable noise levels are in a screening system.

Criteria can be as simple (e.g., presence or not of a pupil reflex) or as complex (e.g., alteration of an EKG) as required. The first step in establishing them should be an evaluation of the performance of test systems that have not been treated (i.e., negative controls). There will be some innate variability in the population, and understanding this variability is essential to selling some *threshold* for *activity* that has an acceptably low level of occurrence in a control population. Figure 3.2 illustrates this approach.

What end points are measured as inputs to an activity criterion are intrinsic in the screen system, but may be either direct (i.e., having some established mechanistic relationship to the end point that is being predicted in man, such as gene mutations as predictive of carcinogenicity in man) or correlative. Correlated variables (such as many of those measured in *in vitro* systems) are "black box" predictors—compounds causing certain changes in these variables have a high probability of having a certain effect in man, though the mechanism (or commonality of mechanism) is not established. There is also, it should be noted, a group of effects seen in animals the relevance of which in man is not known. This illustrates

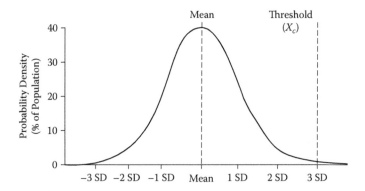

Figure 3.2 Setting thresholds using historical control data. The figure shows a Gaussian (*normal*) distribution of screen parameters; 99.7% of the observations in the population are within three standard deviations (SD) of the historic mean. Here the threshold (i.e., the point at which a datum is outside of *normal*) was set at X_c = mean + 3 SD. Note that such a screen is one-sided.

an important point to consider in the design of a screen—one should have an understanding (in advance) of the actions to be taken given each of the possible outcomes of a screen.

3.6   Analysis of screening data

Screening data present a special case that, due to its inherent characteristics, is not well served by traditional statistical approaches [11,12,19,20].

Why? First consider which factors influence the power of a statistical test. Gad [11] established the basic factors that influence the statistical performance of any bioassay in terms of its sensitivity and error rates. Recently, Healy [21] presented a review of the factors that influence the power of a study (the ability to detect a dose-related effect when it actually exists). In brief, the power of a study depends on seven aspects of study design:

- Sample size
- Background variability (error variance)
- Size of true effect to be detected (i.e., objective of the study)
- Type of significance test
- Significance level
- Decision rule (the number of false positives that one will accept)
- Sensitivity of measurement of variable of interest

There are several ways to increase power—each with a consequence.

Action	Consequence
Increase the sample size	Greater resources required
Design test to detect larger differences	Less useful conclusions
Use a more powerful significance test	Stronger assumptions required
Increase the significance level	Higher statistical false-positive rate
Use one-tailed decision rule	Blind to effects in the opposite direction

Timely and constant incorporation of knowledge of test system characteristics and performance will reduce background variability and allow sharper focus on the actual variable of interest. There are, however, a variety of nontraditional approaches to the analysis of screening data.

3.6.1   Univariate data

3.6.1.1   Control charts
The control chart approach [22], commonly used in manufacturing quality control for screening of defective product units, offers some desirable characteristics.

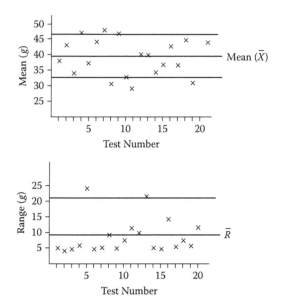

Figure 3.3 Example of a control chart used to *prescreen* data (actually, explore and identify influential variables), from a portion of a functional observational battery. See text discussion for explanation.

By keeping records of cumulative results during the development of screen methodology, an initial estimate of the variability (such as standard deviation) of each assay will be available when full-scale use of the screen starts. The initial estimates can then be revised as more data are generated (i.e., as we become more familiar with the screen).

The following example shows the usefulness of control charts for control measurements in a screening procedure. Our example test for screening potential muscle-strength suppressive agents measures reduction of grip strength by test compounds compared with a control treatment. A control chart was established to monitor the performance of the control agent (1) to establish the mean and variability of the control, and (2) to ensure that the results of the control for a given experiment are within reasonable limits (a validation of the assay procedure).

As in control charts for quality control, the mean and average range of the assay were determined from previous experiments. In this example, the screen had been run 20 times previous to collecting the data shown. These initial data showed a mean grip strength X of 400 g and a mean range R of 90 g. These values were used for the control chart (Figure 3.3). The subgroups are of size 5. The action limits for the mean and range charts were calculated as follows:

$$X \pm 0.58R = 400 \pm 0.58 \times 90 = 348\text{--}452 \text{ (from the X chart)}$$

Then, using the upper limit (*du*) for an n of 5,

$$2.11R = 2.11 \times 90 = 190 \text{ (the upper limit for the range)}$$

Note that the range limit, which actually established a limit for the variability of our data, is, in fact, a *detector* for the presence of outliers (extreme values).

Such charts may also be constructed and used for proportion or count types of data. By constructing such charts for the range of control data, we may then use them as rapid and efficient tools for detecting effects in groups being assessed for that same activity end point.

3.6.1.2 Central tendency plots

The objective behind our analysis of screen data is to have a means of efficiently, rapidly, and objectively identifying those agents that have a reasonable probability of being active. Any materials that we so identify may be further investigated in a more rigorous manner, which will generate data that can be analyzed by traditional means. In other words, we want a method that makes out-of-the-ordinary results stand out. To do this we must first set the limits on *ordinary* (summarize the control case data) and then overlay a scheme that causes those things that are not ordinary to become readily detected (*exposed*, in EDA terms) [23,24]. One can then perform *confirmatory* tests and statistical analysis (using traditional hypothesis testing techniques), if so desired.

If we collect a set of control data on a variable (say scores on our observations of the righting reflex) from some number of *ordinary* animals, we can plot it as a set of two histograms (one for individual animals and the second for the highest total score in each randomly assigned group of five animals), such as those shown in Figure 3.4 (the data for which came from 200 actual control animals).

Such a plot identifies the nature of our data, visually classifying them into those that will not influence our analysis (in the set shown, clearly scores of 0 fit into this category) and those that will critically influence the outcome of an analysis. In so doing, the position of control (*normal*) observations is readily revealed as a *central tendency* in the data (hence the name for this technique).

We can (and should) develop such plots for each of our variables. Simple inspection makes clear that answers having no discriminatory power (0 values in Figure 3.4) do not interest us or influence our identifying of an outlier in a group and should simply be put aside or ignored before continuing on with analysis. This first stage, summarizing the control data, gives us a device for identifying data with discriminatory power (extreme values), thus allowing us to set aside the data without discriminatory power.

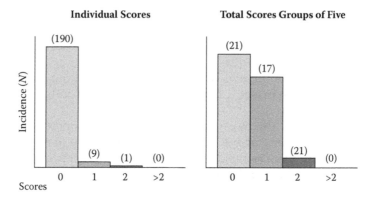

Figure 3.4 Plotting central tendency. Possible individual scores for righting reflexes may range from 0 to 8 [3]. Group total scores would thus range from 0 to 40. (Shown are the number of groups that contain individual scores in the individual categories.)

Focusing our efforts on the remainder, it becomes clear that although the incidence of a single, low, nonzero observation in a group means nothing, total group scores of 2 or more occurred only 5% of the time by chance. So we can simply perform an extreme value screen on our *collapsed* data sets, looking for total group values or individual values that are beyond our acceptance criteria.

The next step in this method is to develop a histogram for each ranked or quantal variable, by both individual and group. *Useless* data (those that will not influence the outcome of the analysis) are then identified and dropped from analysis. Group scores may then be simply evaluated against the baseline histograms to identify those groups with scores divergent enough from control to be either true positives or acceptably low-incidence false positives. Additional control data can continue to be incorporated in such a system over time, both increasing the power of the analysis and providing a check on screen performance.

3.6.2   *Multivariate data*

The traditional acute, subchronic, and chronic toxicity studies performed in rodents and other species also can be considered to constitute multiple end point screens. Although the numerically measured continuous variables (body weight, food consumption, hematology values) generally can be statistically evaluated individually by traditional means, the same concerns of loss of information present in the interrelationship of such variables apply. Generally, traditional multivariate methods are not available, efficient, sensitive, or practical [25].

3.6.2.1   The analog plot

The human eye is extremely good at comparing the size, shape, and color of pictorial symbols [26–30]. Furthermore, it can simultaneously appreciate both the minute detail and the broad pattern.

The simple way of transforming a table of numbers to a sheet of pictures is by using analog plots. Numbers are converted to symbols according to their magnitude. The greater the number, the larger the symbol. Multiple variables can be portrayed as separate columns or as differently shaped or colored symbols [31].

The conversion requires a conversion chart from the magnitude of the number to the symbol size. The conversion function should be monotonic (e.g., dose, and the measured responses should each change in one direction according to a linear, logarithmic, or probit function). Log conversion will give more emphasis to differences at the lower end of the scale, whereas a probit will stabilize the central range of response (16%–84%) of a percentage variable. For example, for numbers x, symbol radium r, and plotting scaling factor k, a log mapping will give:

$$x = 1 \qquad r = k$$
$$x = 10 \qquad r = 2k$$
$$x = 100 \qquad r = 3k$$

To compare different variables on the same sheet requires some form of standardization to put them on the same scale. Also, a choice must be made between displaying the magnitude of the numbers or their significance [32,33]. Two possibilities are—

- Express each mean as a percentage change from a control level or overall mean (*a means plot*).
- Calculate effects for meaningful contrasts (*a contrasts plot*).

The analog plot chart in Figure 3.5 shows relationships for five measures on a time-versus-dose basis, allowing ready evaluation of interrelationships and patterns.

A study using 50 rats of each sex in each of five groups (two controls and three increasing doses) measured body weight and food and liquid consumption every week or month for 2 years. This resulted in 3 variables × 2 sexes × 5 groups × 53 times × 50 animals. Means alone constituted some 1600 four-digit numbers.

Body weight gains from the period immediately preceding each consumption measurement were used, since these were less correlated. For each variable and at each time, the sums of squares for group differences were divided into four meaningful contrasts:

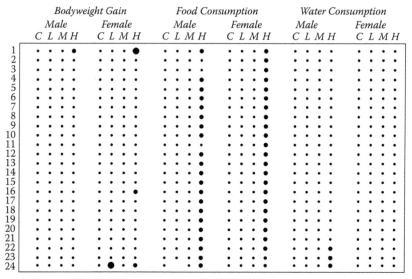

Note: Square root relationship

Key: *C* = control *A* vs *B*. *L* = low vs *A* + *B*. *M* = medium vs *A* + *B* + low. *H* = High vs *A* + *B* + low + medium.

Figure 3.5 Analog plot for dose–response contrasts. One of many possible approaches to graphically presenting multidimensional data. In this case, various effects—day of dosing, dose response, and magnitude of response—are simultaneously portrayed, with the size of each circle being proportional to the magnitude of the measured value.

> Control A vs. control B
> Control A + B vs. low
> Control A + B + low vs. medium
> Control A + B + low + medium vs. high

To make the variables comparable, the sums of squares were standardized by the within-group standard deviations. Contrast involving doses can be compared with the contrast for the difference between the controls, which should be random. The clearest feature is the high-dose effect for food consumption. However, this seems not to be closely correlated with changes in body weight gains. Certain changes can be seen at the later measurement times, probably because of dying animals.

There are numerous approaches to the problem of capturing all the information in a set of multi–end point data. When the data are continuous in nature, approaches such as the analog plot can be used [34,35]. A form of control chart also can be derived for such uses when the goal is detecting effect rather than exploring relationships between variables. When the data are discontinuous, other forms of analysis must be used.

Just as the control chart can be adapted to analyzing attribute data, an analog plot can be adapted. Other methods are also available.

3.7 Study design

Table 1.6 in Chapter 1 introduced the principal components of study design for which some guidance was given in ICH (International Conference on Harmonisation of Technical Requirements for Registration of Pharmaceuticals for Human Use) S7A. Here I would like to offer some practical and more detailed guidance.

3.7.1 Animal model

While the S7A suggests the use of conscious, unrestrained, telemeterized, and/or trained animals, this may not be practical in all instances. Some cardiovascular and respiratory measures may require or be best served by using animals that are restrained but acclimated, for example.

Two equally important points are not addressed in the guidelines. The first is whether the animals employed in *in vivo* studies need to be naïve (implying that any animals are used for evaluating one compound cannot be used for another, and therefore significantly increasing animal usage and costs) or if they could be used repeatedly for evaluations of different compounds (each in a range of doses covering multiple intended therapeutic dose levels but not toxicologic levels, with a suitable washout period between compounds). The latter is preferable unless there is reason to believe (generally based on knowledge of mechanisms or of actions of the class of compounds, as well as observations during a study) that the compound tested irreparably alters the animals employed in its evaluation.

The second question is whether *sensitive* functionally compromised or disease model animals should be employed for some of the safety pharmacology applications, particularly those of the respiratory and renal systems, where the presence of significant functional reserves can serve to reduce the ability of test methods to detect changes. There are at least two good arguments for using sensitive models [36]. The first is that healthy animals and humans possess significant reserve capacities, which would preclude modest but significant effect being detected at doses in and reasonably near the therapeutic range (as evidenced by the ability of individuals to lead quite normal lives with only single lungs or kidneys). The second is that drugs are actually used by sick individuals, and certainly the organ systems involved in the targeted disease claim for a drug will be compromised in those individuals using the drug in the marketplace, and therefore animal models with similarly compromised functions would best serve to identify potential problems in those individuals using the drug therapeutically.

The major argument against such models is a conservative one—that there is inadequate baseline information and experience with such models to reliably evaluate results. This argument is particularly suspect when one considers that the animal models used to identify and verify therapeutic efficacy before going into man are just such diseased or compromised models.

3.7.2   Group size

Experience should rule here as any formal statistical power analysis would require assumptions as to what a meaningful change would be and as to what the variability of a group of animals (fairly well controlled in normal rats, but much less so in dogs or telemetrized and/or trained animals) on study will be.

Generally, groups of four rodents or four nonrodents per data point should be adequate, except in the case of a free-standing Irwin screen or functional observational battery (FOB), where many endpoints are assessed in the same animal (with high variability associated with individual endpoints). Here the animals of one sex would be advisable.

3.7.3   Statistical design

A randomized block design should be the standard in this case, ensuring that adequate washout times are allowed between treatments.

3.7.3.1   Dose levels and test concentrations

Dose levels and test concentrations should be selected according to the following criteria:

- They should define the dose– (concentration–) response curve.
- The time course should be investigated when feasible.
- Doses should include and exceed the primary pharmacodynamic or therapeutic range. In the absence of adverse effects on safety pharmacology parameters, highest tested doses should produce moderate adverse effects in this or other safety assessment studies.
- Some effects in the toxic range (e.g., tremors during EKG recording) may confound the interpretation of safety pharmacology effects and may also limit dose levels.

References

1. Zbinden, G., Elsner, J., and Boelsterli, U.A., Toxicological Screening, *Regul. Toxicol. Pharmacol.*,1984, 4:275–286.
2. Anderson, S., and Hauck, W., A new procedure for testing equivalence in comparative bioavailability and other clinical trials, *Commun. Statist. Thero. Meth.*, 1983, 12:2663–2692.

3. Gad, S.C., A neuromuscular screen for use in industrial toxicology, *J. Toxicol. Environ. Health*, 1982a, 9:691–704.

4. Gad, S.C., Statistical analysis of behavioral toxicology data and studies, *Arch. Toxicol.* (Suppl.), 1982b, 5:256–266.

5. Gad, S.C., *Statistics and Experimental Design for Toxicologists*, 4th Edition. Boca Raton, FL: CRC Press, 2005.

6. Bergman, S.W., and Gittins, J.C., Screening procedures for discovering active compounds. In: *Statistical Methods for Pharmaceutical Research Planning* (Peace, K., ed.), New York: Marcel Dekker, 1985, pp. 359–376.

7. Tukey, J.W., *Exploratory Data Analysis*, Reading, PA: Addison-Wesley, 1977.

8. Redman, C., Screening compounds for clinically active drugs. In: *Statistics in the Pharmaceutical Industry* (Buncher, C.R. and Tsya, J., eds.), New York: Marcel Dekker, 1981, pp. 19–42.

9. Hoaglin, D.C., Mosteller, F., and Tukey, J.W., *Understanding Robust and Exploratory Data Analysis*, New York: John Wiley, 1983.

10. Hoaglin, D.C., Mosteller, F.D., and Tukey, J.W., *Exploring Data Tables, Trends, and Shapes*, New York: John Wiley, 1985.

11. Gad, S.C., An approach to the design and analysis of screening studies in toxicology, *J. Am. Coll. Toxicol.*, 1988, 8:127–138.

12. Gad, S.C., Principles of screening in toxicology with special emphasis on applications to neurotoxicology, *J. Am. Coll. Toxicol.*, 1989, 8:21–27.

13. Garrett, R.A., and London, J.P., *Fundamentals of Naval Operations Analysis*, Annapolis, MD: U.S. Naval Institute, 1970.

14. Martin, Y.C., Kutter, E., and Austel, V., *Modern Drug Research*, New York: Marcel Dekker, 1988, pp. 31–34, 155, 265–269, 314–318.

15. Pennie, W.D., Use of cDNA microassays to probe and understand the toxicological consequences of altered gene expression, *Toxicol. Lett.*, 2000, 112–113:473–477.

16. Nuwaysir, E.F., et al., Microassays and toxicology: the advent of toxicogenomics, *Mol. Carcinog.*, 1999, 24:153–159.

17. Page, N.P., Concepts of a bioassay program in environmental carcinogenesis. In: *Environmental Cancer* (Kraybill, H.F., and Mehlman, M.A., eds.), New York: Hemisphere Publishing, 1977, pp. 87–171.

18. Bickis, M.G., Experimental design. In: *Handbook of In Vivo Toxicity Testing* (Arnold, D.L., Grice, H.C., and Krewski, D.R., eds.), San Diego: Academic Press, 1990, pp. 128–134.

19. Gad, S.C., Screens in neurotoxicity: Objectives, design, and analysis, with the observational battery as a case example, *J. Am. Coll. Toxicol.*, 1989, 8:1–18.

20. Gad, S.C., Statistical analysis of screening studies in toxicology with special emphasis on neurotoxicology, *J. Amer. Coll. Toxicol.*, 1989, 8:171–183.

21. Healy, G.F., Power calculations in toxicology. *A.T.L.A.*, 1987, 15:132–139.

22. Montgomery, D.C., *Introduction to Statistical Quality Control*, New York: John Wiley and Sons, 1985.

23. Velleman, P.F., and Hoaglin, D.C., *Applications, Basics, and Computing of Exploratory Data Analysis*, Boston: Duxbury Press, 1981.

24. Tufte, E.R., *The Visual Display of Quantitative Information*, Cheshire, CT: Graphic Press, 1983.

25. Young, F.W., Multidimensional scaling. In: *Encyclopedia of Statistical Sciences*, vol. 5 (Katz, S., and Johnson, N.L., eds.), New York: John Wiley, 1985, pp. 649–659.

26. Anderson, E., A semigraphical method for the analysis of complex problems, 1960, *Technometrics* 2:387–391.
27. Andrews, D.F., Plots of high dimensional data, *Biometrics*, 1972, 28:125–136.
28. Davison, M.L., *Multidimensional Scaling*, New York: John Wiley, 1983.
29. Schmid, C.F., *Statistical Graphics*, New York: John Wiley, 1983.
30. Cleveland, W.S., and McGill, R., Graphical perception and graphical methods for analyzing scientific data, *Science*, 1985, 229:828–833.
31. Wilk, M.B., and Gnanadesikan, R., Probability plotting methods for the analysis of data, *Biometrics*, 1986, 55:1–17.
32. Kruskal, J.B., Multidimensional scaling by optimizing goodness of fit to a nonmetric hypothesis, *Psychometrika*, 1964, 29:1–27.
33. Kass, G.V., An exploratory technique for investigating large quantities of categorical data, *Appl. Stat.*, 1980, 29:119–127.
34. Chernoff, H., The use of faces to represent points in K-dimensional space graphically, *J. Am. Stat. Assoc.*, 1973, 68:361–368.
35. Chambers, J.M., et al., *Graphical Methods for Data Analysis*, Boston: Duxbury Press, 1983.
36. Milano, S., Introducing disease animal models in safety pharmacology: a way to reevaluate the risk, *MDS Safety Pharmacology Symposium*, Lyon, France: MDA Pharma, December 5–6, 2002.

chapter four

Cardiovascular system

4.1   Introduction

The cardiovascular system is one of the three principal vital organ systems whose functions have to be evaluated during safety pharmacology studies. Cardiovascular system functioning is maintained by cardiac electrical activity and by pump-muscle function, which contribute to hemodynamic efficacy. The aim of cardiovascular safety pharmacology is to evaluate the effects of test substances on the most pertinent components of this system so that potentially undesirable effects can be detected before engaging in clinical trials [1–3] and to predict the potential risk of such events in long-term clinical usage. In the basic program, a detailed hemodynamic evaluation is carried out in unanesthetized dogs, primates, or pigs with implanted telemetry sensors. It includes cardiac electrophysiology and vascular pressure evaluations to assess both arrhythmogenic and other cardiovascular risk. An *in vitro* assessment of effects on potassium channel transport (the human *ether-a-go-go*-related gene, or hERG, assay) is also performed to complete the evaluation. The basic program can be preceded by rapid and simple testing procedures during the early drug discovery stage. It should be completed, if necessary, by specific supplementary studies, depending on the data obtained during the early clinical trials.

4.2   History

4.2.1   Special case (and concern)—QT prolongation

Drugs that alter ventricular repolarization (generally characterized as drugs that *prolong the QT interval*) have been associated with malignant ventricular arrhythmias (especially the distinctive polymorphic ventricular tachycardia called *torsades de pointes*, TdP) and death [4–8]. Many of the drugs now known to alter ventricular repolarization were developed as antiarrhythmics (e.g., dofetelide, sotalol), but others (e.g., cisapride, terfenadine) were developed without the expectation of any effect upon electrically excitable membranes. This has lead to the International Conference on Harmonisation of Technical Requirements for Registration of Pharmaceuticals for Human Use (ICH) promulgating ICH S7A [9] with specific guidance for evaluation of such electrophysiologic function.

Table 4.1 Drugs Causing QT Prolongation and *Torsades de Pointes* [18,19, 45]

Agent	Drugs
Class IA antiarrhythmic agents	Quinidine, disopyramide, procainamide
Class III antiarrhythmic agents	Amiodarone, dofetilide, D-sotalol, dibutilide
Calcium antagonists	Bepridil, terodiline
Antihypertensives	Ketanserin (α-adrenoceptor antagonist)
Antidepressant agents	Amitryptiline, citalopram, clomipramine, desipramine, doxepin, imipramine, maproptiline, nortriptiline, zimelidine
Antifungal agents	Fluconazole, itraconazole, ketoconazole, micoconazole
Antihistamine agents	Astemizole, diphenhydramine, hydroxizine, terfenadine
Anticancer agents	Amsacrine, doxorubicine, zorubicine
Antimicrobial agents	Amantadine, amphotericin, clarithromycin, clindamycin, cotrimoxazole, erythromycin, grepafloxacin, pentamidine, sparfloxacin, spiramycin, trimethoprin-sulphamethoxazole, troleandomycin
Antimalarial agents	Chloroquine, halofantrine, quinine
Antipsychotic agents	Chlorpromazine, haloperidol, flufenazine, lithium, mesoridazine, pimozide, prochlorperazine, roperidol, sultopride, sertindole, risperidone, thioridazine, trifluoperazine
Miscellaneous agents	Cisapride, probucol, indapamide, K^+ wasting diuretics (e.g., furosemide)

Prolongation of the cardiac action potential duration (APD), reflected in the electrocardiogram (ECG) as QT interval prolongation (QT prolongation), is associated with potentially lethal polymorphic ventricular tachyarrhythmia (TdP) [10–12]. QT prolongation may occur spontaneously in subjects having mutations in genes encoding for potassium or sodium channels regulating normal cardiac repolarization (congenital long QT syndromes, LQTS) [13,14]. By the mid-1990s QT prolongation was also associated with a growing number of drugs (antiarrhythmic as well as non-antiarrhythmic) and drug combinations (acquired LQTS). Many (if not most) of these drugs prolong QT by inhibiting the rapid component of the delayed rectifying potassium current, I_{Kr}; the a-subunit through which this current flows is encoded by the hERG [15,16]. The likelihood of developing TdP is considered to be related to the magnitude of prolongation of ventricular repolarization [17]. Non-antiarrhythmic drugs that affect QT at therapeutic doses typically show less than 10–15 msec mean QT prolongation (Table 4.1). The frequency with which these drugs evoke TdP is very

Table 4.2 Cardiovascular System Safety Pharmacology Evaluations

Core

- Hemodynamics (blood pressure, heart rate)
- Autonomic function (cardiovascular challenge)
- Electrophysiology (EKG in dog)

QT prolongation (noncore)

An additional guideline, ICH S7B, addresses the assessment of potential for QT prolongation. In the meantime, CPMP 986/96 indicates the following preclinical studies should be conducted prior to first administration to man:

- Cardiac action potential *in vitro*
- ECG (QT measurements) in a cardiovascular study that would be covered in the core battery
- hERG channel interactions (hERG expressed in HEK 293 cells)

Source: Committee for Proprietary Medicinal Products (CPMP), Points to consider: The assessment of the potential for QT interval prolongation by non-cardiovascular medicinal products, *986/96 guidance,* 1997. [46]

low, estimated in the range of 1 in 2000 to 20,000+ patients, dependent on other risk factors in the treated population; hence, treatment-evoked TdP is unlikely to be detected within the context of a traditional clinical development program. Thus there is an imperative to develop techniques and strategies to detect potential for TdP hazard early in development, ideally, prior to first exposure in man (FIM).

4.2.1.1   Regulatory developments

The first regulatory body to issue comments on specific concerns related to the potential for drugs to induce QT interval prolongation was the Committee for Proprietary Medicinal Products (CPMP) 986/96 guidance [46] (Table 4.2). CPMP identified specific methodologies for both preclinical and clinical studies, including stimulation frequencies for *in vitro* preparations, numbers of ECG leads and chart speed, and QT interval correction (QTc) strategies for heart rate changes. Over the following decade plus, the discussion has evolved, as other regulatory bodies have provided guidance (Table 4.3) and we have accumulated data and understanding. The current S7A and S7B guidance describes in detail study design consideration for these evaluations, including a strong preference for *in vivo* ECG studies conducted in conscious, instrumented, and unstressed animal models. Arguably, the QT prolongation issue made safety pharmacology a regulatory subdiscipline in 2000–2001.

The objective of preclinical cardiovascular safety testing is to identify either (1) EKG/ECG signals of potential TdP hazard, or (2) a change in heart rate (pulse) and blood pressure indicators for purposes of evaluating the potential risk to the first humans to be exposed to a new drug, and for

Table 4.3 Current International Guidelines and Draft Documents
on Preclinical Assessment of TdP Hazard

Document	Date	Comment
Ministry of Health and Welfare, Japan: *Notes on Applications for Approval to Manufacture (Import) New Drugs*, issued in 1975	1975	Requested evaluation of: "…effects of the test substance on the…central nervous system, peripheral nervous system, sensory organs, respiratory and cardiovascular systems, smooth muscles including uterus, peripheral organs,…renal function,…and adverse effects observed in clinical studies" [p.71].
Japanese Guidelines for Nonclinical Studies of Drugs Manual, 1995. Yakuji Nippo, Limited, Tokyo, 1995.	1995	Studies in Lists A and B became de facto international blueprints for general/safety pharmacology evaluations until issue of ICH S7A. "Normally, anesthetized animals are used"; hence, anesthetized animal preparation (and particularly the barbiturate-anesthetized dog) became the standard for cardiovascular-respiratory evaluations [p.128].
Committee for Proprietary Medicinal Products (CPMP EU). *Points to Consider: The Assessment of the Potential for QT Interval Prolongation by Noncardiovascular Medicinal Products* (http://www.emea.eu.int/ pdfs/human/ swp/09869en.pdf)	1997	First regulatory document addressing the TdP hazard with pharmaceuticals; credited with generating academic and interindustrial cooperation to share existing data and to generate collaborative efforts to rapidly produce realistic experimental and clinical approaches for identification of preclinical signals a TdP hazard.
U.S. Department of Health and Human Services, Food and Drug Administration, Center for Drug Evaluation and Research, Center for Biologics Evaluation and Research, ICH: *Guidance for Industry. S7A Safety Pharmacology Studies for Human Pharmaceuticals*, July 2001 (http://www.ifpma.org/ich5s. html)	2001	Provides the general study design framework for *in vitro* and *in vivo* preclinical evaluations of TdP. Specifically places evaluations addressing risk for repolarization-associated ventricular tachy-arrhythmia within the safety pharmacology domain.

Table 4.3 (Continued) Current International Guidelines and Draft Documents on Preclinical Assessment of TdP Hazard

Document	Date	Comment
Therapeutic Products Directorate Guidance Document (Canada): *Assessment of the QT Prolongation Potential of Non-Antiarrhythmic Drugs*, 2001 (http://www.hc-sc. gc.ca/hpb-dgps/therapeut/ htmleng/guidmain.html)	2001	States explicitly that development of a non-antiarrhythmic drug with a preclinical signal (*in vitro* or *in vivo*) of TdP hazard "should be pursued only if it is expected to provide a major benefit for a serious disease or disorder for which safer alternatives are not available, or if the cardiotoxicity is attributable to a metabolite generated in animals, but not in humans."
ICH *Guideline on Safety Pharmacology Studies for Assessing the Potential for Delayed Ventricular Repolarization (QT Interval Prolongation) by Human Pharmaceuticals (S7B), Step 2,* 2002.	2002	Presents a tiered testing scheme recommending *in vitro* ion current and repolarization evaluations and an *in vivo* QT assessment in appropriate species; provides a current assessment of the pros and cons of available techniques while recognizing that this area is in great flux and recommending that new technologies be evaluated and applied as they become available.
FDA: *The clinical evaluation of QT/ QTc interval prolongation and proarrhythmic potential for non-antiarrhythmic drugs.* FDA preliminary concept paper. Nov. 15, 2002	2002	Accepts ICH S7B guidelines, when finalized, of preclinical assessment of potential TdP hazard; provides a starting point for discussion as to how to address preclinical signals within subsequent clinical development.

stipulating additional precautions and safeguards to protect those individuals while effects of a new drug on human ventricular repolarization are being established. The preclinical testing scheme is well developed in ICH S7B, with general aspects of study design referenced to ICH S7A. The assessment of risks with the data generated was not so well delineated initially, but has evolved (as will be discussed). An integrative, tiered physiological approach is currently recommended with evaluations directed at the subcellular, cell, tissue, and intact organism levels (Figure 4.1), should signals of interest be detected. Compounds can be examined for effects at a target receptor by evaluating effects on specific ion currents in heterologous expression systems, such as the hERG, which is stably expressed

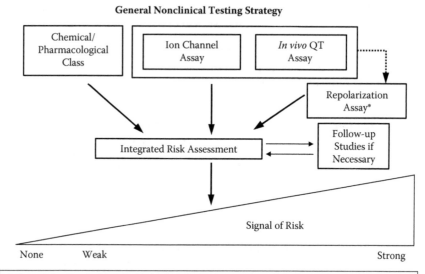

Figure 4.1 General nonclinical testing strategy.

in a mammalian cell line. hERG is a logical target receptor. Most drugs that prolong QT interval and elicit TdP in humans also block the hERG-related I_{Kr} current; however, hERG is not the only target potentially affecting ventricular repolarization. Alteration in the function of IKs (KVLQT1 and MinK) or Ina (hH1) may potentially also produce repolarization abnormalities that would predispose toward TdP (though no drug moieties have actually been so identified). Intact ventricular myocyte, isolated tissues (Purkinje fibers), and isolated hearts represent the successive levels of increased integration; cellular systems allow the study of effects of a compound on all currents during the action potential in aggregate; tissue and organ systems permit addition of cell–cell interactions and interstitial environmental effects. Finally, studies using intact (preferably conscious, unstressed) animals permit evaluation of pharmacodynamic and tissue distribution effects, homeostatic compensatory effects, and effects of metabolites. Surface electrocardiography results in intact animals also provide the bridge into the initial clinical ECG program by providing direct assessments of effects on QT interval and other aspects of ECG morphology (T and U waves, arrhythmias). Hence, it is critical that the intact animal study designs model the intended FIM exposure. In this regard, while there is intense interest in preclinical arrhythmia models with which to evaluate directly treatment-evoked arrhythmia potential, there is currently no consensus that any of the available models are relevant for predicting TdP hazard in humans.

Remarkable progress has been made toward a rational consensus approach for preclinical evaluation of TdP hazard. Because the exposures in these isolated cell, tissue, and organ systems may approach the solubility limits of test substances and be many-fold greater than therapeutic exposures, guidance as to appropriate margins (now provided in the form of the cardiac safety index, or CSI) needed to be developed; with recent pronouncements from the Food and Drug Administration (FDA) figures and experts suggest that "weak" or negative early findings of QT interval prolongation [20] should not trigger a monitoring burden in later clinical development. There are several limitations for *in vitro* studies of putative target molecules. A heterologously expressed channel is not in its native cellular environment; this may have effects on basal function and, even more insidiously, may selectively impact drug action. Such studies do not permit the examination of drug effects in diseased or compromised tissues, nor do they permit assessment of effects peculiar to chronic exposure (e.g., up-regulation or down-regulation). The effects of major human metabolites of the parent drug can only be performed if the metabolites are known. In studies of a single target channel (e.g., hERG), effects of the agent on other ion targets will always be missed. The disadvantages of native cell studies include increased expense and time. The process of isolating cells may change the function of the channel itself and possibly the response to the drug. Microenviroments of cells in artificial culture medium may be different from those in intact tissues and magnify or mask responses. Typically, available human cells have been isolated from diseased hearts that have been exposed to elaborate, uncontrolled drug regimens, further complicating the interpretation of the results. Rats (and probably mice) are known to lack the I_{Kr} current thought to be responsible for much of the acquired LQTS in humans; hence, ECG evaluations should not be conducted in these species. On the other hand, anecdotal reports of compound-evoked TdP in rats cause concern that ion currents/combinations other than I_{Kr} can also produce TdP, and suggest a role for rodents ECGs in safety evaluation.

In intact animals, much (perhaps too much) attention has been paid to appropriate strategies to "correct" animal QT intervals for underlying changes in heart rate [21]. In an ideal setting, baseline and postdrug ECGs would be collected at identical heart rates; practically, it is nearly impossible to ensure similar heart rates when only small amounts of ECG data (<1 min) can be collected during each study interval. Bazett recognized this problem in the early 1920s and modeled human QT and relative rate (RR) interval data to produce a QT value corrected to 60 bpm (QTc Bazett) [22]. However, Bazett himself recognized the weaknesses of a mathematically modeled *pseudo-physiological* parameter and recommended against the widespread use of his QTc formula. QTc Bazett is well known not to model QT-RR relationships usefully in dogs, and its use has led to misidentification of cardiotoxic effects of drugs (especially those that produce direct or indirect effects on heart rate) in this species [23,24]. With automated data collection and wireless ECG telemetry systems, it is possible to collect vast amounts of ECG data from experimental animals and to develop baseline and postdrug QT-RR relationships in individual animals, obviating the need for QT correction and its inherent assumptions. In the above, several strategies are being considered to genereate physiological heart rate ranges for QT-RR relationships. The simplest is to monitor ECGs over 24-hr periods taking advantage of circadian heart rate variation [25].

No single preclinical effect (marker) is yet considered reliable to predict TdP hazard in humans. A blocking effect at the level of hERG *in vitro* does not necessarily translate to an effect on APD or QT in the same (let alone another) species and under other experimental or clinical conditions. Similarly, an effect on APD and QT in animals does not necessarily accurately predict TdP in humans due to species differences in cardiac depolarization/repolarization and in metabolite formation [17]; and the lack of ion current, ADP, or QT signals from preclinical testing does not predict that a compound will be free from such effects in humans. Hence, an integration of signals of unwanted effects in the chain of events from blockade of specific ion currents to generation of actual arrhythmias must be continuously evaluated across the spectrum of preclinical designs, throughout clinical development in humans, and in the postdevelopment drug surveillance. Ultimately, the human risk of TdP is identified through careful assessment of ventricular repolarization and adverse events in relevant clinical populations [26]. Current draft guidance states that data on potential QT changes must accompany the safety summary of the regulatory submission from (in aggregate) at least 100 volunteers (preferably including both males and females) on relevant doses of drug and suitable number on placebo.

But with growing frequency preclinical signals are being detected for almost all moieties at some concentration from the required testing (particularly the hERG), which presents a risk assessment problem. If all moieties provoke some signal, what signals should be ignored? Such signals provoke decisions about (1) additional preclinical investigation, (2) additional clinical safety monitoring and investigations of treatment-evoked TdP hazard in humans, and (3) continuing, delaying, or even discontinuing development. Thus, for preclinical signals of TdP risk, the following should be considered:

- *Detection and evaluation of signals.* Efforts continue to better identify and understand the significance of signals from *in vitro* and *in vivo* studies; hypothetically, two categories of signals and associated levels of concern are defined [27].

- A *strong signal* is an observation for which a current majority (consensus) opinion supports a linkage to a possible TdP risk. A strong signal is a definite cause for concern. Examples of strong signals could include treatment-evoked ventricular arrhythmias (except occasional supraventricular or ventricular actopics) in animals or humans, or association with a chemical or pharmacological class of agents known or suspected to pose a risk of TdP.
- A *weak signal* is an observation supporting a linkage to a possible TdP risk, but for which there is no consensus opinion. A weak signal is a possible cause for concern. Examples of weak signals could include (a) an average increase in QT interval (compared with baseline or placebo) outside the normal range, (b) significant change from baseline in rate-corrected QT interval (QTc, using a species-appropriate method), (c) treatment associated QT and/or QTc absolute values outside of normal ranges identified for specific species, and (d) signals from any individual preclinical *in vitro* (hERG channel, APD, etc.) or *in vivo* (ECG evaluation in animals or a pathophysiologic animal model), with the exception that certain treatment-evoked arrhythmias constitute weak signals because of current uncertainties in extrapolation of nonclinical findings to humans.

It is conceivable that two or more weak signals could, in aggregate, constitute a strong signal of TdP risk based upon the consensus criteria. Alternatively, if neither the preclinical testing nor the early clinical testing show any electrophysiological effects related to delayed repolarization (e.g. signals), the likelihood of the new active substance showing important proarrhythmic effects during its clinical use is considered remote. In evaluating signals of TdP risk, consideration is given to the following:

- *Magnitude*: Is the magnitude sufficient to be discerned from the background variability of the model being used?
- *Dose–response*: Are dose– or concentration–response relationships apparent in the data? Evidence of dose–response can help distinguish treatment-evoked effects from experimental variation.
- *Reversibility and reproducibility*: Is the effect reversible upon removal of exposure; can the effect be reproduced following re-exposure (e.g., treatment evoked)?
- *Cardiac safety index (CSI)*: What is the ratio of the exposure (dose or concentration) where a signal is first apparent (lowest effect level, LOEL), compared with a projection or measured efficacious unbound plasma concentration (ED90, EC90) associated with an efficacious therapeutic dose? Although ability to extrapolate nonclinical effects to humans is imprecise, a large CSI may provide a basis for reducing concern whereas a CSI ≤10 may increase concern about a particular

signal. Several groups are developing data supporting validation of therapeutic index (TI) calculations based upon signals generated in hERG expression systems and human TdP risk.

Another form of this is the calculation of a CSI, or ratio between the level where there are clear therapeutic effects (ED90) and that at which there is an indication of minimal hERG activity (such as an EC10). A CSI of 30 is generally considered an absolute minimum safety margin. A CSI of 100 is a more common margin [28].

- *Tissue distribution:* Is the drug accumulated or sequestered in cardiac tissues? Drugs that accumulate in cardiac tissue may impart risk that is not reflected by either total or unbound circulating drug concentrations. Drugs that prolong ventricular repolarization in a fashion unrelated to its pharmacokinetic profile or appear at the nadir of drug concentrations should be evaluated for tissue accumulation or for the presence of long-lived metabolites.

- *Metabolites:* With regard to major metabolites, it is important to distinguish whether these are human specific or present in sufficient concentrations in humans to pose an arrhythmia risk.

- *Species specificity:* Is a signal demonstrable in two or more species? The absence of a signal in multiple species at similar exposures may lessen concern that a particular signal will be observed in man or indicate the presence of a species-specific active metabolite.

- *Direct and indirect effects on QT interval:* When QT or QTc interval prolongation or alterations in T-wave morphology are noted in a multiple nonclinical dose study, consideration should be given to whether these findings are a result of direct effects on cardiac conduction or an indirect effect related to toxicity. Consideration should be given to whether concurrent with the changes in ventricular repolarization, there are cardiac lesion or alterations in clinical chemistry parameters (e.g., hypokalemia, hypomagnesemia, hypocalcemia). This distinction is important in recognizing the primary toxicity produced by a drug and monitoring for this event in clinical investigations (e.g., the appropriate choice of biomarker).

In the absence of a validated or generally accepted marker of TdP hazard, differentiation of strong and weak signals of TdP risk is significant due to impact on the clinical program. If a weak signal is present, the clinical program is spurred to more fully explore the potential for TdP risk in volunteers and in patients with additional risk factors, while perhaps initially increasing the intensity of safety monitoring procedures. The primary objective of these clinical evaluations is relatively straightforward: detect/define the extent of treatment-evoked prolongation of ventricular repolarization (if any) in the relevant patient populations. However, if a strong signal is present, the clinical program (1) must increase the intensity of safety monitoring, (2) must define the relationship between the signal and the TdP risk in the intended patient population, and (3) may elect to restrict study entry criteria to exclude patients with additional risk factors (e.g., long basal QT). The clinical challenge is daunting: to prove that the frequency of an already-low-probability adverse event (TdP) is not unduly increased in patients exposed to treatment. Only intensive clinical investigation can reduce TdP concern generated by preclinical signals, and it is imperative that this be done prior to initiation of the pivotal efficacy (Phase III) program, so that safety monitoring and entry criteria are the least restrictive, consistent with patient safety and good clinical practice [29].

Preclinical aspects of detection and evaluation of signals of TdP hazard have evolved rapidly since the early 1990s and are now codified in the ICH Safety Pharmacology guidance (S7B). The S7B is a unique document that envisions additional refinements in strategy and technology and encourages their use; however, operationally it is not clear how tripartite consensus will be maintained as preclinical testing evolves with time in areas not specifically addressed in S7B—there is no regulatory precedent. Progress on preclinical markers of TdP also illustrates the challenge to develop specific *biomarkers* of efficacy and toxicity end points identified in animals and "validated" for drugs entering clinical development. Given the knowledge developed over the past century on the physiology and pathophysiology of the heartbeat, and the intense academic and industrial cooperation since CPMP 1997 (Table 4.3), it is sobering that as of yet no single biomarker of TdP hazard is yet established from preclinical testing that is uniformly viewed as a predictive surrogate for TdP hazard in humans. QT prolongation appears to be a mechanistically understood candidate marker, common to and easily monitored in both animals and humans, but the debate continues as to thresholds of prolongation past which the risk of TdP is actually increased in either animals or humans [30]. Rather, there is an expanding list of signals of potential TdP hazard emanating from both preclinical and clinical studies that must be integrated and interpreted for decision making. The late Dr. Gerhard Zwinden, an early champion of organ function as toxicological targets during an era dominated by morphologic toxicology, is vindicated; today's ideal non-antiarrhythmic drugs should not influence IKr, APD, and QT, all organ function targets.

The association between abnormalities of repolarization and life-threatening arrhythmias is stronger than some other associations between laboratory abnormalities and clinical events. For example, there are drugs (tacrine) and inborn errors of metabolism (Gilbert's syndrome) that cause wild excursions in liver-function tests, but with no adverse consequences. In contrast, although the severity of proarrhythmia at a given QT duration varies from drug to drug and from patient to patient, no drug is known to alter ventricular repolarization without inducing arrhythmias,* and each of the several congenital long-QT syndromes is associated with an elevated incidence of malignant arrythmias.

With any given repolarization-altering drug, the risk of malignant arrhythmia seems to increase with increasing QT interval, but there is no well-established threshold duration below which a prolonged QT interval is known to be harmless. The extent of QT prolongation seen with a given drug and patient may be nonlinearly related to patient factors (e.g., sex, electrolyte levels, and others) and to serum levels of the drug and/or its metabolites. The actual incidence of malignant arrhythmias, even in association with the drugs most known to induce them, is relatively low, so failure to observe malignant arrhythmias during clinical trials of ordinary size and duration does not provide substantial reassurance.

Abnormal repolarization and the associated arrhythmias are the end results of a causative chain that starts with alternations in the channels of ionic flux through cell membranes. Some cells (e.g., those of the Purkinje system or midmyocardium) seem especially susceptible to these changes. At a substrate level, the links on the chain are alterations in the time course of the action potential, alterations in the propagation of action potentials within a given cell, and alterations in the propagation of action potentials from cell to cell within syncitia and from tissue to tissue within the heart. At a higher level of aggregation, one sees *afterdepolarizations* in the terminal portion of the action potential, spontaneous beats triggered by afterdepolarizations, propagation of these beats to other cells, and re-entrant excitation.

With these considerations in mind, the problem of altered repolarization should be integrated into drug development by:

- *In vitro* screening of the drug and its metabolites for effects on ion channels (especially the I_{Kr} calcium channel)
- *In vitro* screening of the drug and its metabolites for effects on action-potential duration
- Screening of the drug and its metabolites for altered repolarization in animal models
- Focused preclinical studies for proarrhythmia if altered repolarization is seen in preclinical screening or in patients

---

* Some QT-prolonging drugs (e.g., amiodarone [31]) are not reported to have caused many arrhythmic deaths, but this observation must be interpreted carefully. In a population with a high incidence of life-threatening arrhythmias, a drug with both proarrhythmic and antiarrhythmic effects might cause a net reduction in arrhythmias, and the arrhythmias that it induced might not be attributed to it. In a population whose native arrhythmias were not life threatening, the same drug might result in a net decrease in mortality.

Some specific techniques that can be employed include patch-clamp studies and hERG protein expression system.

4.2.2   Patch-clamp studies using recombinant cells expressing hERG channels

Most pharmaceuticals associated with *torsades de pointes* inhibit rapidly delayed rectifier current, I_{Kr}. Therefore particular attention to assays for I_{Kr} is prudent for assessing risk of QT interval prolongation [32].

Using the voltage-clamp technique, outward or inward ionic currents can be measured from single cell preparations. Because of inherent difficulties associated with recording I_{Kr} in native myocytes, much of the available data for this current has been obtained using recombinant cell lines expressing *h*ERG.

Inhibition of other outward (repolarizing) currents (e.g., I_{to}, I_{Kl}, I_{Ks}) or increase in inward (depolarizing) ionic currents (e.g., I_{Na}) could also lead to QT interval prolongation and, therefore, should be considered when investigating the mechanism(s) for QT interval prolongation [33–35].

4.2.3   hERG protein expression system

When transfected with hERG alone or hERG in association with genes for potential regulating subunits (e.g., MiRP1), appropriate cell systems express a K^+ channel that displays biophysical and pharmacological properties similar to I_{Kr}. Several expression systems have been used to test the activity of test substances on hERG current.

The cell line should be selected based upon appropriate levels of *h*ERG expression.

The cell line should be tested at appropriate intervals to confirm the stability of the current from the expressed channel.

When in the presence of endogenous currents, the presence or absence of subunits (directly associated regulatory proteins) and kinases or phosphatases controlling regulatory phosphorylation sites can affect the pharmacology of the expressed channel protein relative to native cardiac ion channels.

Mammalian (e.g., CHO, HEK-293, COS-7) rather than *Xenopus* oocytes should be selected for expressing hERG since a major limitation of *Xenopus* model is that test substances can accumulate in the oocyte yolk, resulting in significant variability and error in potency estimates.

hERG rather than I_{Kr} present murine atrial tumor AT-1 or HL-1 cells derived from transgenic mice should be preferred. "The channels expressed by these cells are similar to neonatal/fetal mouse cardiac myocytes raising some concerns about de-differentiation in these

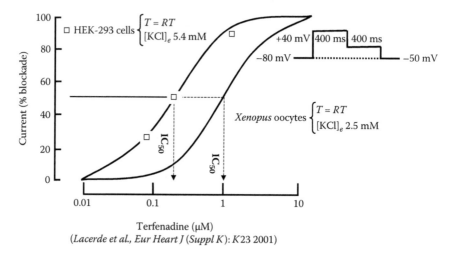

Effects of Terfenadine on *h*ERG Channels Expressed in HEK-293 Cells and *Xenopus* Oocytes

Terfenadine (μM)
(*Lacerde et al., Eur Heart J (Suppl K): K23* 2001)

Figure 4.2 Effects of terfenadine on *h*ERG channels expressed in HEK-293 cells and *Xenopus* oocytes.

tumor-derived cells" [34]. Test results obtained are also clearly dependent on culture temperature (see Table 4.4), and such must be clearly specified with any provided or published data.

4.2.3.1   Relevance of hERG to QT prolongation

Compounds that are associated with adverse drug reactions (ADRs) of QT prolongation, arrhythmias such as TdP, and sudden death predominantly have a secondary pharmacological interaction with the rapidly activating delayed rectifier potassium channel I_{Kr}. The gene encoding this channel has been identified as hERG. Testing of compounds for interactions with the hERG channel allows the identification of potential risk of QT prolongation in humans and can be used as a screen in development candidate selection.

4.2.3.2   Expression and recording systems

HEK-293 cells have been transfected with cDNA for hERG-1 to produce a stable expression system (the cell line was developed in the lab of Dr. Craig January at the University of Wisconsin).

The relevance and relative importance of these measurements in conjunction with all of the other components of cardiovascular safety assessment is still not clear. Certainly it is influential in making drug development decisions [21] but should not be a critical determining factor on its own.

Table 4.4 IC_{50} of Several Drugs against hERG Channel
(HEK-293) Conveyed Current Determined at 22°C and 35°C

Drug	IC_{50} (µM) at 22°C	IC_{50} (µM) at 35°C
Terfanadine	0.02	0.3
Loratadine	4.7	7.9
Dephenidramine	3.9	2.7
Fexofenadine	11	13.6
Cisapride	0.03	0.04
Sotalol	810	270
Erythromycin	3158	200

4.2.4 Cardiovascular function testing

Traditionally, safety pharmacology screens for cardiovascular function have been performed in the anesthetized dog or rat. In the rat, diastolic, systolic, and mean arterial blood pressures (DAP, SAP, MAP) are recorded via direct cannulation of a major artery (e.g., femoral or carotid), and the test article is injected directly into a vein while the animal remains anesthetized. Heart rate is determined electronically from the pressure signal, and other physiological parameters may be monitored, such as body temperature, respiration rate, and ECG. Often, an escalating dose strategy is employed to determine the minimum dose necessary to see any changes in cardiovascular function.

Cardiovascular monitoring in dogs remains the preclinical gold standard [36,37] and is usually performed with a thermistor-tipped Swan-Ganz–type catheter advanced directly into the pulmonary artery for measurement of cardiac output (CO), pulmonary arterial pressure (PAP), and core body temperature. A second catheter is placed in the caudal vena cava for measurement of central venous pressure (CVP) and for collection of blood for hematology and clinical chemistry. An additional catheter can be advanced into the aorta for measurement of DAP, SAP, and MAP and for collection of arterial blood for blood gas analysis. A catheter also can be advanced to the left ventricle for measurement of lateral ventricular pressure (LVP) and dp/dt, a measurement of cardiac contractility. Simultaneously, a catheter may be placed in the bladder for continual voiding during surgery and testing, and the urine may be collected for measurements of volume, electrolyte concentration, and protein content or for determination of creatinine clearance. Again, as in the experiments in rats, ECG leads are used to record gross changes in cardiac electrophysiology in anesthetized dogs. This setup permits the simultaneous monitoring of a wide variety of important cardiovascular parameters. The obvious drawback is that these heavily instrumented animals are subject to surgery beforehand (with a recovery period before use in a study) and significant movement limitations to permit accurate recording of these multiple parameters.

The potential interaction of the anesthetic compound or the anesthetized state with the test article is generally ignored. The rationale for this is that safety pharmacology screens are by definition meant to identify gross physiologic changes in response to doses well above the expected therapeutic levels. However, if the anesthetic increases the response threshold to a test article, then we may be underestimating the safety index for that compound. If some parameters of cardiovascular function are depressed by the anesthetic, or if there are unidentified interactions of the test article with the anesthetic, we hope that these can be unmasked by escalating the dose off the compound being tested. Traditional toxicity testing attempts to avoid complex drug interactions such as those with an anesthetic.

For basic cardiovascular monitoring in rats, many researchers have switched to tailcuff sphygmomanometry as a noninvasive technique for obtaining DAP, SAP, MAP, and heart rate in awake, nonanesthetized animals. The main drawback in using the tailcuff to measure blood pressure is the need for regular restraint whenever you want to take a reading. Ideally, rats must be trained beforehand to be comfortable and remain docile in the restraining apparatus. The stress of restraint is well documented to significantly alter physiological state and, in particular, cardiovascular function. In addition, most tailcuff systems work best when the rat is warmer than standard room temperature, requiring the use of heat lamps or other methods to increase local body temperature in the tail. Heating may impact the cardiovascular recording and the functional state of the animal. Cuff sphygmomanometry also can be used with dogs (tail or leg) and monkeys (arm), but these animals must be acclimated for these conscious, restrained blood pressure measurements. Another drawback of this technique is the discontinuity of measuring cardiac function. The rats, dogs, and monkeys can monitored only while restrained and in intervening times must be allowed to move freely. Nevertheless, cuff sphygmomanometry (tail or limb) is a valuable addition and noninvasive alternative allowing for the measurement of basic cardiovascular parameters in nonanesthetized animals.

More recently, researchers have been using radiotelemetric implants to record a number of cardiovascular parameters in awake, freely moving animals [38]. Most often, the lower abdominal aorta is cannulated and connected to a transmitter sutured directly to the wall of the peritoneal cavity. After surgery, the animal is allowed to recover from the anesthesia and returned to its home cage. A receiver unit is placed next to the cage to pick up the radio signals emitted by the telemetry device. Measurements of blood pressure, heart rate, body temperature, and ECGs are routinely taken this way. The major advantage of radiotelemetry is the elimination of external influences, because all recordings can be performed from a site remote or hidden from the test animal, and the animal can remain in its home cage throughout the measurement and testing procedures. Of course, drug administration requires a separate setup, instrumentation, or restraint, depending on the route of delivery. In some instances, drug administration can be done from a remote site, too. Telemetry has the potential to record the purest data on the effects of a test article on cardiovascular function, because stress from external events and handling are greatly reduced for an animal being studied with telemetry system.

An important advantage of radiotelemetry over cuff sphygmomanometry is the ability to record data continuously. Following implantation, the physiological detectors are always in place, and data storage space is the only practical limitation on data collection throughout an experiment. The major drawback of radiotelemetry is the expense of the transmitters and their maintenance. The greater expense leads many researchers to do long-term studies, either testing one compound over a very long time (e.g., chronic toxicity testing) or testing multiple compounds with adequate washout periods between them. Of course, with novel test articles, one is never sure of the adequacy of a washout period, particularly if the tissue distribution and half-life information for these novel compounds are not known. If multiple compounds are tested in one animal, then one cannot perform histopathology or tissue distribution evaluations at the same time. Currently with telemetry, one cannot measure all of the parameters that can be monitored via direct cannulation in anesthetized animals. Advances in electronics and miniaturization engineering may change this in the near future.

Radiotelemetry devices are also subject to interference from other nearby electronic instruments, and they have a tendency to display a baseline drift over time. Careful attention to the study setup and design as well as improved sensor design can minimize these problems. To compensate for drift one may need to periodically *calibrate* the animal and telemetry system with responses to known agents, but this is not desirable when testing an unknown compound. Lastly, maintenance of a telemetry-instrumented colony can be expensive. Many researchers find that the cost of telemetry implants is not feasible for short-term evaluations. In spite of these drawbacks, radiotelemetry may produce the most accurate picture of the effects of a novel compound on cardiovascular function.

As imaging technology has improved (particularly with improved image resolution) and its costs have declined, nonclinical safety evaluation of both drugs and medical devices (particularly devices intended for cardiovascular uses) have begun to utilize imaging technologies for evaluating the safety of new therapeutics [39,40].

Magnetic resonance imaging (MRI) relies mainly on the detection of hydrogen nuclei in water and fat to construct high-resolution images. The contrast in these images results from different tissue environments. MRI is based on the same principles as liquid-state nuclear magnetic resonance (NMR), that is, the behavior of nuclei in a magnetic field under the influence of radio frequency (RF) pulses; but the hardware, pulse sequences, and data processing are somewhat different. Improvements in electronics and computers since the 1990s have given MRI resolution capabilities in intact organisms down to approximately 3mm, and therefore tremendous potential as a tool for studying mechanisms of toxicology. MRI techniques provide detailed information on the response of specific organs to toxicants and can also be used to monitor xenobiotic metabolism *in vivo*.

Echocardiography (ECHO) is the use of ultrasound technology to access various aspects of cardiac function and morphology. By using different ECHO windows (standard placements of the ECHO probe) and types of ultrasound propagation, qualitative information can be obtained regarding indices of cardiac size, systolic function, diastolic function, and hemodynamics. Furthermore, certain parameters are a reflection of an integrated input of cardiac functions and can be used for a global impression of cardiac function [41].

In the present studies, the parameters measured varied because of differences in heart size, anatomy, species movement artifacts, and equipment used. However, certain parameters are constant across the studies, and these are used for comparative purposes. Two of these parameters (or surrogate) are also those used in the *Common Terminology Criteria for Adverse Events* (CTCAE), version 3 (v3), an adverse event rating, which is used for assessing adverse events during clinical trials. These parameters include an assessment of left ventricular ejection fraction and fractional shortening of the left ventricle.

Table 4.5 presents a glossary of the terms used, how these reflect indices of cardiac function, and how changes in these parameters reflect decreases in cardiac function.

Clinical studies in oncology tumor vascular disrupting agents have shown indications of cardiac effects, including QTc interval prolongation, measurable plasma troponin-T levels, cardiac hypoxia, and cardiac ischemia. GLP IND-enabling (investigational device exemption) studies of several agents have indicated minimal-slight cardiopathology in the dog and rat. Following such observations, the cardiotoxic potentials have been investigated by means of ECHO, together with histiopathy and cardiac biochemistry, in dog, pig, and cynomolgus monkey.

Results with studies suggest that the use of ECHO can demonstrate changes in cardiac function at lower doses that can predict more significant changes at higher doses.

Studies on cardiac function in the dog and monkey have shown neither troponin-T nor the cardiac selective isoforms of creatine kinase that appeared to be reliable indicators of cardiac pathology of systolic dysfunction.

Table 4.5 Glossary of Echocardiographic Terms and Changes
Associated with a Decreased Cardiac Function

Heart size

Decrease in parameters is associated with a decrease in cardiac function, as
there is a reduced volume.
- LVIDd: left ventricular inner diameter at diastole (mm)
- LAD: left atrial diameter (mm)

Systolic function

FxS: fractional change in left ventricular diameter (%)
- A decrease is associated with decreased cardiac function, reflecting
reduced systolic contraction.

FxArea: fractional change in left ventricular area (%)
- A decrease is associated with decreased cardiac function, reflecting a reduced
area of systolic contraction.

LVIDs: left ventricular inner diameter (mm)
- An increase is associated with decreased cardiac function, reflecting
reduced systolic contraction.

PEP/LVET: ratio of the pre-ejection period to left ventricular ejection time
- An increase is associated with deficits in contractibility or high afterload, a
relatively shortened period of systolic contraction.

VCFm: Mean velocity of circumferential fiber shortening (circ/sec)
- A decrease is associated with decreased cardiac function, reflecting a
reduced rate of systolic contraction or reduced contractility.

Diastolic function

IVRT: left ventricular isovolumic relaxation time (ms)
- An increase is associated with impaired myocardial relaxation.

Hemodynamics

Heart Rate (bpm)

VRI: velocity time integral of aortic flow profile (mm)
- Related to stroke volume
- Decreased with a decrease in cardiac function

VTI x HR: proportional to cardiac output (mm/min)
- Decreased with a decrease in cardiac function

AoVel: peak aortic velocity (mm/sec)
- Decreased with a decrease in cardiac function

Global function

EPSS: separation of the mitral leaflets from the septal wall during the early
wave of mitral flow (mm)
- Increased with a decrease in cardiac function; may be associated with mitral
regurgitation

Histiopathic examination of the heart is necessarily an integral part of the echocardiographic studies performed to investigate effects on cardiac function in the dog and monkey. Properly formulated patterns of histio-pathologic findings similar to those reports in the safety and toxicology studies, with myocardial necrosis of the papillary muscle, intraventricular septum, and left ventricular wall, are seen.

Minimally cardiotoxic Cmax values for such agents appear to vary considerably between species and also with the use of different formu-lations. A further confounding factor to consider is the variability in sampling times between studies, a critical aspect when dealing with an intravenously administered compound with a short half-life (2 h) as is the case with many protein therapeutic agents. Results with studies sug-gest that the use of ECHO can demonstrate changes in cardiac function at lower doses that can be predictive of more significant changes at higher doses.

4.2.5   Conscious rodent, dog, and primate telemetry studies

Effects on blood pressure, heart rate, lead II ECG, core body temperature, and locomotor activity are now most commonly conducted using animals with implanted telemetry devices (such as those from DataSciences) in rats, guinea pigs, dogs, pigs, or primates. Effects on behavior can be cap-tured on video using closed-circuit television (CCTV) for dog and pri-mate studies. Repeated administration and interaction studies can be performed.

4.2.5.1   Six-lead ECG measurement in the conscious dog

Conscious studies using integrated telemetry systems devices for mea-surement of blood pressure and six chest lead ECG measurements (V2, V4, V6, V10, rV2, and rV4) can be routinely performed. ECG interval analysis is performed on the V2 lead (RR, PR, QT, QTc intervals, QRS duration). QT dispersion can also be measured. Locomotor activity can be monitored and behavior captured on video using CCTV.

- In addition to validated systems for automatic measurement of ECG parameters, ECGs can be reviewed by veterinary cardiology services to detect any transparent abnormalities.
- Colonies of telemetered animals can be set up and maintained for repeat use.
- Respiration rate measurements can be taken from dogs in slings using a pneumograph system.
- An animal-specific correction of QT interval can also be derived for each dog or primate based on individual variability of QT interval with rate using the Framingham equation.

Studies to assess the effects of compound and any known metabolites on ECG and cardiac action potentials are recommended. Changes in action potential duration and other parameters measured are a functional consequence of effects on the ion channels, which contribute to the action potential. This *in vitro* test is considered to provide a reliable risk assessment of the potential for a compound to prolong QT interval in man.

4.2.6 Systems for recording cardiac action potentials

These include a range of currently available methodologies, some of which can be incorporated into existing study designs.

- Isolated ventricular Purkinje fibers from dog or sheep
- Isolated right ventricular papillary muscle from guinea pig
- Continuous intracellular recording of action potentials and on-line analysis of resting membrane potential, maximum rate of depolarization, upstroke amplitude, and action potential duration using Notocord HEM data acquisition system
- Assessment of use-dependent and inverse use-dependent actions by stimulation at normal, bradycardic, and tachycardic frequencies (e.g., see below inverse use-dependent properties of sotalol in dog Purkinje fibers)

4.2.6.1 Cloned human potassium channels

Assessment of effects on cloned hERG K channels with stable expression in a cell line by measurement of whole cell K current (I_{Kr}) using voltage clamp is the most frequent approach. Other cloned human ion channels (e.g., KvLQT1/minK-I_{Ks} currents) are also possible.

4.2.6.2 Cardiac action potential in vitro—Purkinje fibers

Intracellular recording of action potentials from cardiac Purkinje fibers isolated from dog or sheep ventricle are most common. Measurement of the maximum rate of depolarization and action potential duration to detect sodium and potassium channel interactions respectively [42] is thus measured.

4.2.6.3 Monophasic action potential in anesthetized guinea pigs

Epicardia monophasic action potential recording using suction/contact pressure electrodes is possible according to Carlsson et al. [7], allowing simultaneous measurement of ECG.

4.2.6.4 ECG by telemetry in conscious guinea pigs

Lead II ECG recording is performed using DataSciences' telemetry device. Repeated administration and interaction studies can be performed in this paradigm.

4.2.6.5   *Hemodynamics and ECG in anesthetized or conscious dogs or primates*

- Conscious studies using DataSciences' telemetry for blood pressure and lead II ECG or the ITS system for blood pressure and six chest lead ECG measurements (including QT dispersion)
- Anesthetized studies using MI² data capture system with additional measurement of blood flow in selected vascular beds, cardiac output, respiratory, and left ventricular function

4.3   Summary

Understanding and adequately and properly evaluating the adverse cardiovascular pharmacological effects of potential new drugs represents the visible expression of the most compelling current area of development and controversy in drug safety evaluation and drug development. As we have improved our knowledge and tools, however, it has become clear that the emphasis on ECG effects alone have diverted attention from other aspects of cardiovascular function that are just as important for identifying and understanding potential cardiovascular risk [43,44]. Of particular concern is detection and appreciation of the significance hemodynamic change and alteration in cardiac contractibility (inotropy).

The current situation remains analogous to that of genotoxicity testing in the 1970s, however. Many test systems are being employed without an absolute ability to use the resulting data to evaluate risk. The challenge lies in understanding all of the information that is available and that we can collect, and integrating it into a risk evaluation model that both protects patients and does not inadvertently block or retard the advancement into clinical use those therapeutics that are safe and medically valuable.

References

1. Lacroix, P., and Provost, D., Basic safety pharmacology: the cardiovascular system, *Therapie*, 2000, 55:63–69.
2. MacKenzie, I., Safety pharmacology requirements for the development of human cardiac/cardiovascular pharmaceuticals, *Drug Devel. Res.*, 2002, 55: 73–78.
3. Malik, M., and Camm, A.J., Evaluation of drug-induced QT interval prolongation: implications for drug approval and labeling, *Drug Safety*, 2001, 24:323–351.
4. Cashin, C.H., Dawson, W., and Kitchen, E.A., The pharmacology of benoxaprofen (2-[4-chlorophynl]-α-methyl-5-benzoxazole acetic acid), LRCL 3794, a new compound with anti-inflammatory activity apparently unrelated to inhibition of prostaglandin synthesis, *J. Pharm. Pharmacol.*, 1977, 29:330–336.
5. Graf, E., et al., Animal experiments on the safety pharmacology of lofexidine. *Arzneim.-Forsch./Drug Res.*, 1982, 32(II)(8a):931–940.

6. Bramm, E., Binderup, L., and Arrigoni-Martelli, E., An unusual profile of activity of a new basic anti-inflammatory drug, timegadine, *Agents Actions,* 1981, 11:402–409.

7. Carlsson, L., et al., Electrophysiological characterization of the prokinetic agents cisapride and mosapride *in vivo* and *in vitro*: implications for proarrhythmic potential, *J. of Pharmacol. and Ex. Therapeutics,* 1997, 282:220–227.

8. Takasuna, K., et al., General pharmacology of the new quinolone anti-bacterial agent levofloxacin, *Arzneim.-Forsch./Drug Res,* 1992, 42(I)(3a):408–418.

9. ICH S7A, *Safety Pharmacology Studies for Assessing the Potential for Delayed Ventricular Repolarization (QT Interval Prolongation) by Human Pharmaceuticals,* 2002.

10. Thomas, S.H.L., Drugs, QT interval abnormalities and ventricular arrhythmias. *Adverse Drug React. Toxicol.,* 1994, Rev 13:77–102.

11. De Ponti, F., Poluzzi, E., and Mantanaro, N., Organising evidence on QT prolongation and occurrence of torsades de pointes with non-antiarrhythmic drugs: a call for consensus, *Eur. J. Clin. Pharmaco.,* 2001, 57:185–209.

12. Moss, A.J., The QT interval and torsade de pointes, *Drug Safety,* 1999, 21:5–10.

13. Ackerman, M.J., The long QT syndrome: ion channel diseases of the heart, *Mayo Clin. Proc.,* 1998, 73:250–269.

14. Viskin, S., Long QT syndromes and torsade de pointes, *Lancet,* 1999, 354:1625–1633.

15. Haverkamp, W., et al., The potential for QT prolongation and proarrhythmia by non-antiarrhythmic drugs: clinical and regulatory implications, *Cardiovasc. Res.,* 2000, 47:219–233.

16. Anderson, M.E., et al., Cardiac repolarization: Current knowledge, critical gaps, and new approaches to drug development and patient management. *AHJ,* 2002, 144:769–781.

17. Hammond, T.G., et al., Methods of collecting and evaluating non-clinical electrophysiology data in the pharmaceutical industry: results of an international survey, *Cardiovascular Res.,* 2001, 49:741–750.

18. Altman, E.M., *Cardiovascular Therapeutic,* 2nd ed., Philadelphia: W.B. Saunders Company, 2002.

19. Hardman, J.C., and Limbird, L.E., *Goodman & Gilman's The Pharmacological Basis of Therapeutics,* 9th ed., New York: McGraw-Hill, 1996.

20. Anon, QT internal early negative signal would lesson burden in later clinical trial, *The Pink Sheet,* January 27, 2003, pp. 13–14.

21. Bonate, P., and Russell, T., Assessment of QTc prolongation for non-cardiac-related drugs from a drug development perspective, *J. Clin. Pharmacol.,* 1999, 39:349–358.

22. Bazett, H.C., An analysis of time: relations of electro-cardiograms, *Heart,* 1920, 7:353–370.

23. Turner, R.A., *Screening Methods in Pharmacology,* vols. I and II, New York: Academic, 1965, pp. 42–47, 60–68, 27–128.

24. Domer, F.R., *Animal Experiments in Pharmacological Analysis,* Springfield, IL: Charles C. Thomas, 1971, pp. 98, 115, 155, 164, 220.

25. Walker, M.J.A., and Pugsley, M.K., *Methods in Cardiac Electrophysiology,* Boca Raton, FL: CRC Press, 1998.

26. Temple, R., Are surrogate markers adequate to assess cardiovascular disease drugs? *J. of the American Med. Assoc.,* 1999, 282:790–795.

27. Kinter, L.B., and Valentin, J.P., Safety pharmacology and risk assessment, *Fundamental & Clinical Pharmacology*, 2002, 16:175–182.
28. Gad, S.C., Nonclinical safety assessment of the cardiovascular toxicity of drugs and combination medical devices. In: *Target Organ Toxicity: The Cardiovascular System* (D. Acosta, ed.), Philadelphia: Taylor & Francis, 2008, pp. 33–58.
29. Joy, J.P., et al., Prediction of torsade de pointes from the QT interval: analysis of a case series of amisulpride overdoses, *Clinical Pharmacology & Therapeutics* 2011, 90:243–245.
30. Rosendorff, C. *Essential Cardiology: Principles and Practice*, Philadelphia: W.B. Saunders Company, 2001
31. Hohnloser, S.H., Klingenheben, T., and Singh, B.N., Amiodarone-associated proarrhythmic effects, *Ann. Intern. Med*, 1994, 121:529–535.
32. Cavero, I., and Crumb, W.J., Mechanism-designed assessment of cardiac electrophysiology safety of pharmaceuticals using human cardiac ion channels, *Business Briefing: Pharma. Tec.*, July 2001, pp. 1–9.
33. Lacerda, A.E., et al., Comparison of block among cloned cardiac potassium channels by non-antiarrhythmic drugs, *Eur. Heart J.*, 2001, Suppl. K:K23–K30.
34. Rampe, D., et al., A mechanism for the proarrhythmic effects of cisapride (Propulsid): high affinity blockade of the human cardiac potassium channel hERG, *FEBS Letters*, 1997, 417:28–32.
35. Fenichel, R.P., and Koerner, J., Development of drugs that alter ventricular repolarization, Internet-based draft, 1999.
36. Detweiler, D.K., Electrocardiography in toxicologic studies. In: *Comprehensive Toxicology* (Sipes, I.G., McQueen, C.A., and Gadndolfi, J.A., eds.), New York: Pergamon Press, 1997, pp. 95–114.
37. Osborne, B.E., and Leach, G.H.D., The beagle electrocardiogram, *Fd. Cosmet. Toxicol.*, 1971, 9:857–864.
38. Abernathy, F.W., Flemming, C.D., and Sonntag, W.B., Measurement of cardiovascular response in male Sprague-Dawley rats using radiotelemetric implants and tailcuff sphymomanometry: a comparative study, *Toxicol. Methods*, 1995, 5:89–98.
39. Webster, J.G., *Medical Instrumentation*. New York: John Wiley & Sons, 1998.
40. Riccardi, M.J., Beohar, N., and Davidson, C.J., Coronary catheterization and coronary angiography. In: *Essential Cardiology* (Rosendorf, C., ed.), Philadelphia: W.B. Saunders, pp. 204–226.
41. Kaddoura, S. *ECHO Made Easy*. London: Churchill Livingstone, 2002.
42. Burns, B.D., and Paton, W.D.M., Depolarization of the motor end-plate by decamethonium and acetylcholine, *J. Physiol*, London, 1951, 115:41–73.
43. Sarazan, R.D., et al., Cardiovascular function in nonclinical drug safety assessment: current issues and opportunities, *Int. J. of Toxicol.*, 2011, 30:272–286.
44. Laverty, H.G., et al., How can we improve our understanding of cardiovascular safety liabilities to develop safer medicines? *Brit. J. of Pharmacol.* 2011, 163:675–693.
45. Lipicky, R.J., Drug induced torsade de pointes and implications for drug development, *J. of Cardio. Electrophys*, 2004, 15:475–495.
46. Committee for Proprietary Medicinal Products (CPMP), Points to consider: The assessment of the potential for QT interval prolongation by non-cardiovascular medicinal products, *986/96 guidance*, 1997.

chapter five

Central nervous system

Many drugs (Table 5.1) have both intentional and unintended central nervous system (CNS) pharmacologic effects (see Table 5.2 for examples).

One characteristic that distinguishes CNS safety pharmacology from discovery or efficacy pharmacology (primary pharmacology) is that CNS safety pharmacology is generally conducted in normal animals. The aim is to see whether the new drug induces adverse changes in normal function, and not whether the drug can have potential therapeutic effects on abnormal function. Not to be included in CNS safety pharmacology studies therefore would be models of pathology or disease (e.g., depression, anxiety), unless it was expected that the therapeutic use of the drug might present particular problems in a specific group of patients. Indeed, the notion of safety might differ radically depending on the intended therapeutic application. For example, when developing an N-methyl-D-aspartic acid (NMDA) antagonist for acute stroke, it would not be a very important safety factor if the drug induced a reversible psychotomimetic state, providing the patient's life was saved. On the other hand, if such a drug were to be used preventively in patients suffering repeated minor strokes, the presence or absence of psychotomimetic potential could become a crucial safety consideration.

A further characteristic of CNS safety pharmacology is that studies are almost exclusively carried out *in vivo* and must use conscious animals. In contrast to other organ systems, such as the cardiovascular system, CNS safety cannot rely on *in vitro* techniques, because the neurobiological mechanisms of CNS function are less well understood.

Considering these points, what kinds of studies do come within the scope of CNS safety pharmacology? This topic was extensively developed in the Japanese Guidelines [1], which divided recommended CNS safety pharmacology studies into two categories, A and B. Category A included so-called core battery studies and look more at the indicators that generally are observational and do not require instrumentation: for example, general behavioral observation, measures of spontaneous motor activity, general anesthetic effects and eventual synergism/antagonism with general anesthetics, effects on convulsions (proconvulsant activity and synergy with convulsive agents), analgesia, and body temperature. Category B included effects of the test substance on the electroencephalogram (EEG), the spinal reflex, conditioned avoidance response, and locomotor

Table 5.1 Drugs Having Clinical CNS Side Effects

Therapeutic class	Drugs	Side effect
Beta Blockers	propranolol, metoprolol	lethargy, sleep disturbances
Antihistamines	diphenhydramine, promethazine	sedation
Benzodiazepines	diazepam	sedation, ataxia, amnesia
Opioid Analgesics	morphine	dependence, euphoria, sedation
Antibacterials	ciprofloxacin	dizziness, nausea, insomnia
Cytotoxics	cisplatin	cognitive, neurological dysfunction

Table 5.2 Some Commonly Observed CNS Side Effects

- Amnesia
- Abuse liability
- Anorexia
- Auditory dysfunction
- Cognitive impairment
- Convulsion
- Depression
- Disorientation
- Dizziness
- Drowsiness
- Hallucinations
- Hyperphagia
- Insomnia
- Involuntary movement
- Lethargy
- Motor incoordination
- Nausea
- Personality changes
- Sedation
- Seizures
- Sexual dysfunction
- Tremor
- Visual disturbance

coordination. The governing notion of Category A was that the studies mentioned were obligatory (called "Core Battery" under the International Conference on Harmonisation of Technical Requirements for Registration of Pharmaceuticals for Human Use, or ICH), whereas those mentioned

Table 5.3 Central Nervous System Safety Pharmacology Evaluation

---

- Irwin test*: General assessment of effects on gross behavior and physiological state
- Locomotor activity**: Specific test for sedative, excitatory effects of compounds
- Rotarod: Test of motor coordination
- Anesthetic interactions: Test for central interaction with barbiturates
- Anti-/proconvulsant activity: Potentiation or inhibition of effects of pentylenetetrazole
- Tail flick**: Tests for modulation of nociception (also hot plate, Randall Selitto, tail pinch)
- Body temperature**: Measurement of effects on thermoregulation
- Autonomic function: Interaction with autonomic neurotransmitters *in vitro* or *in vivo*
- Drug dependency: Test for physical dependence, tolerance, and substitution potential
- Learning and memory: Measurement of learning ability and cognitive function in rats

---

* Usually an FOB is integrated into rodent (rat) repeat dose toxicity studies to meet this requirement.
** Properly designed and executed/FOB can also meet their requirements.

in Category B were to be carried out when necessary, but the concept of necessity remained undefined. More recently, the European Agency for the Evaluation of Medicinal Products has proposed a new set of guidelines, ICH S7A [2] in force since June 2001, which have also been adopted in the United States and Japan. These guidelines (Table 5.3) are much less specific than the Japanese Guidelines and include as core battery CNS studies—motor activity, behavioral changes, coordination, sensory/motor reflex responses and body temperature—with the remark that "the central nervous system should be assessed appropriately" [2]. Follow-up studies should include behavioral pharmacology examinations and others. In general, core battery studies should be carried out prior to first administration in humans, whereas the follow-up studies should be carried out prior to product approval. Such core battery studies should be carried out in full accordance with Good Laboratory Practice (GLP), whereas follow-up studies, because of their unique characteristics, require only assurance of data quality and integrity. Comparison of the Japanese Guidelines and current ICH S7A (and all other ICH S guidances) suggests a clear intent by the European authorities to free safety pharmacology from the constraints of a cookbook or checklist approach. On the other hand, the general terms (as with all the other portions of safety pharmacology guidelines except for those governing QTc internal prolongation) do not provide a clear and precise idea of what could or should be done.

The study designs presented here should serve to evaluate the CNS safety pharmacology profile of new entities and excipients while also fulfilling the requirements of the ICH S7A guidelines. The aim, however, remains to provide concrete examples of how established techniques in CNS pharmacology can give body to the bare outlines represented by ICH S7A. After the presentation of the core battery tests, the follow-up procedures within the different domains are presented in order of their methodological complexity that can determine their priority of execution in the drug development process [2].

Historically, there are four broad classes of approaches to assessing nervous system effects of drugs in animals. We shall consider those (behavioral and observational) that address core battery requirements first.

5.1 Core battery CNS procedures

Core battery CNS procedures are by design simple tests, using very traditional techniques, which can be carried out rapidly in a routine and objective fashion. They are the first techniques to be employed in safety assessment and are frequently applied at the very beginning of the discovery process as a screen to eliminate substances with a potential for CNS risk. Because of their use early in the safety evaluation process, such studies are conducted most usually in the rat or the mouse, though (as we shall see) analogous methods have been developed and validated and are in use with dogs, primates, and pigs.

The tests described below generally cover the topics included in the ICH S7A guidelines. Three further tests are described, which are not mentioned in ICH S7A but were included in the Japanese Guidelines: the proconvulsive activity, pain sensitivity, and interaction with barbituarates. It seems, to the present author, that detection of eventual proconvulsant activity or hyperanalgesic effects are topics of eminent concern for CNS safety. As for interaction studies with barbiturates, although hardly novel, these provide a very simple and sensitive means for unmasking eventual sedative or anti-sleep effects that are not always detectable using simple observation tests such as the Irwin or functional observational battery (FOB) [10].

5.1.1 General behavioral observation

A first approach to assessing the global behavioral profile of a novel substance could be with the primary observation procedure in rodents. This approach was originally described by Irwin [3] or the FOB originally developed by Gad [4] and subsequently refined by Haggerty [5], Mattson [6], Moser [7], Moscardo et al. [8], and Redfern et al. [9]. Both tests are mentioned in ICH S7A, but the FOB is more specifically used for assessing adverse nervous system effects. Subsequently Gad, Gad, and Gad [11] described adaptation of the method to dogs, and Gauvin and Baird [12], in primates.

According to the basic Irwin procedure, rats or mice are given the test substance and are repeatedly observed over a 2–3 h period followed by daily observations up to 72 h using a standardized observation grid containing most or all of the following items: mortality, sedation, excitation, sterotypes, aggressiveness, reaction to touch, pain sensitivity, muscle relaxation, loss of righting reflex, changes in gait and respiration, catalepsy, ptosis, corneal reflex, pupil diameter, and rectal temperature. Animals are usually given a relatively high dose first and then the other doses are selected on the basis of the effects observed. The aim is to establish, by means of a limited number of doses, the highest dose that can be given without inducing observable effects, the pharmacologic dose response range, and the first lethal dose. This test also permits an estimate of the duration of activity and the spectrum of behavioral/neurologic effects observed [13].

5.1.2   Functional observational battery

Neurobehavioral evaluations are an important component of testing for the neurotoxic potential of chemicals. Observations made during standard toxicity studies or specialized neurotoxicity studies can provide information important for identifying and/or characterizing neurotoxic effects. A protocol that includes a framework for the systematic recording of observations and manipulations, such as an FOB, is an integral part of neurobehavioral screening. A neurobehavioral test battery can be composed of a variety of endpoints, usually chosen to assess an array of neurological functions, including autonomic, neuromuscular, sensory, and excitability.

The protocols in this unit are divided into (1) observational assessments, and (2) manipulative tests. Each protocol is further subdivided into specific tests or endpoints (Table 5.4). These various end points may be combined into a battery of tests for neurobehavioral screening. Most or all of these protocols or end points should be used in the context of a broad neurobehavioral test battery, whereas judicious selection of specific end points may be appropriate for more focused neurological testing. Originally developed by Gad [4], Moser [7], and Moscardo [8], the functional observational battery provides excellent detailed descriptions of the method using rats, the most common species used.

Note: Generally, a sample size of five animals per sex per treatment is sufficient. It is essential that control animals be treated exactly the same as the other groups except that they are administered only vehicle or formulation (minus drug).

Table 5.4 Endpoints in a Neurobehavioral Screening Battery

Observational assessments	Manipulative tests

Activity levels:
  Home-cage observations
  Open-field observations
  Rearing
	Neurological reflexes/reactions:
	  Pupil response
	  Palpebral reflex
	  Pinna reflex
	  Extensor thrust reflex

Reactivity/excitability:
  Reactivity
  Arousal
	Neuromuscular tests and postural reactions:
	  Grips strength
	  Landing food splay
	  Hopping righting reaction

Gait and postural characteristics:
  Gait descriptions
  Postural descriptions
	Sensory responses:
	  Visual test: approach response
	  Visual test: visual placing
	  Somatosensory tests: touch response
	  Auditory test: click response
	  Nociceptive test: tail/toe pinch
	  Nociceptive test: flexor reflex
	  Proprioceptive positioning test
	  Olfactory test

Involuntary/abnormal motor movements:
  Tremors
  Fasciculations
  Clonus
  Tonus
  Sterotypy
  Bizarre behaviors	

Clinical signs:
  Lacrimation
  Salivation
  Hair coat
  Palpebral closure
  Ocular abnormalities
  Muscle tone/mass	

5.2   Rat

5.2.1   Observational assessments

This section provides observational assessments used to characterize neurological functions, a description of the behaviors being observed, and possible ranking scales to be used, where applicable. The observed behaviors are innate, that is, they do not need to be taught or shaped. Thus, observations require little or no interaction between the observer and the subject, with the possible exception of holding the rat. Because the observer makes judgments regarding these behaviors, the assessments are subjective.

The rat (or other species) may be observed briefly in the home cage, but such observations are quite constrained and therefore limited. The home cage may prevent clear observations due to a variety of factors including available light, position of the cage, and clarity of the cage material. For these reasons, most of these evaluations are made in an open field or arena. The observer should be positioned so as to have a clear, unobstructed view as the rat moves about. Choose an open field that is large enough for the rat to explore, has a nonslippery surface, and has a raised border to prevent the subject from escaping or falling over the edge. Typical open fields include the top of a laboratory cart with a rim or a benchtop with metal sides enclosing the area. Cover the open field with clean absorbent paper that can be changed after each rat (or clean the area after each rat) to eliminate interfering olfactory cues. Environmental conditions (e.g., lighting and temperature) should be held constant from day to day.

The end points described here may be evaluated simultaneously by a single observer during the handling phase and open-field observation period. It is suggested that the open-field period be sufficiently long for the observer to score the animal on all the associated end points (e.g., 2 or 3 min).

5.2.2   Locomotor activity

Locomotor activity can be quantified in rodents by a variety of means (e.g., interruptions of photoelectric beams, activity wheels, changes in electromagnetic fields, Doppler effects, video-image analysis, telemetry) [14,15]. Animals are administered the test substance and are placed in standardized enclosures for a limited observation period (10–30 min) a fixed time after administration of the test substance. In contrast to the Irwin test, which is labor intensive, most activity tests are automated to permit fairly large numbers of animals to be tested simultaneously. As the behavior measured occurs spontaneously, particular care has to be taken to ensure constant experimental conditions (e.g., time of day, temperature, noise level, lighting, apparatus cleanliness) to obtain reproducible results. Although different authors quantify different aspects of locomotion (e.g., small displacements, large displacements, rearing), the basic information yielded by activity meter tests is whether a test substance increases or decreases locomotion. Furthermore, data obtained should be correlated with that obtained from direct observation (Irwin test or FOB) to ensure that apparent decreases are not due to motor incapacity or even to overexcitation.

5.2.3   Motor coordination

Motor coordination is most commonly assessed using a rotarod [16,17]. Rats or mice are placed onto a rod rotating either at a fixed speed or at a constantly increasing speed. The time taken for the animal to fall off the rod, or the number of animals remaining on the rod over a set duration, is measured. To decrease test variability, the animals can be given prior habituation to the rotarod before receiving the test substance. Usually, several animals are tested simultaneously on the same rod, separated physically and visually by partitions. As such, the test lends itself readily to automation.

In contrast to most of the other core battery tests, the rotarod is unidirectional, detecting only the capacity of substances to decrease neuromuscular coordination. On the other hand, when used in conjunction with locomotor activity tests, it provides a useful quantification of the margin of safety between doses of test substances that alter spontaneous activity and those that disturb motor function.

5.2.4   Pain sensitivity

Nociception pain perception can be measured in rodents by a variety of procedures whereby aversive stimulation is applied externally either by heat or electrical stimulation to the tail or paws (tail flick, hot plate, plantar test) or internally by injection into the peritoneum of chemicals (acetic acid, phenylbenzoquinone). Inflammation pain can also be induced by injecting chemicals (carrageenan, formalin) into the paws. More complex approaches induce neuropathic pain by surgical lesions, usually to the sciatic or spinal nerve. For safety pharmacology purposes, usually only the simpler procedures are used. For example, with the hot plate test [18] the animal is placed onto a heated metal plate (54°C) within a vertical cylinder, and the latency to licking its front paws is measured over a short period.

Analgesic effects are clearly demonstrated with major analgesics, such as morphine, at doses that do not change spontaneous locomotion, but false positives can also be obtained with any drug that inhibits locomotion unless this activity is quantified separately. A more interesting application of pain sensitivity tests is for detecting increased pain sensitivity. Drug-induced hyperalgesia could constitute an important risk factor for novel substances.

5.2.5   Convulsive threshold

Convulsions can be induced in rodents by electric shock applied directly across the cerebrum (ECS), by chemical agents administered peripherally, or even by exposure to noise in specific strains of animals. In safety pharmacology it is particularly important to detect proconvulsant activity. On the other hand, although anticonvulsant activity does not in itself constitute a risk, many substances with anticonvulsant activity, for example, benzodiazepines, induce sedation and memory impairment, which have obvious implications for CNS safety. Both kinds of activity can be seen with chemically induced convulsions such as with pentylenetetrazole [16], where a shorter or longer latency to convulsions and deaths can be observed with pro- and anticonvulsants, respectively. A variant of the electroconvulsive method, whereby successive animals are subjected to increasing or decreasing intensities of ECS depending on the occurrence of convulsions with the preceding shock level, permits a sensitive measure of the convulsive threshold, which can vary in both directions and is therefore particularly useful for safety pharmacology purposes.

5.3   Dog

Young (4 to 6 months of age) beagle dogs from laboratory breeders/suppliers should be used. Groups of four or five animals each should be singly housed and provided with water (*ad libitum*) and 1.5 kg of standard laboratory chow in two feedings a day in morning and afternoon. Light should be maintained in a 12 hour light–dark cycle, and dosing to be performed in the morning.

5.3.1   Neurobehavioral screen

An early assumption in developing this screen was that a starting point should be a methodology familiar to laboratory animal veterinarians. If as many as possible of the actual evaluations incorporated into the screen were adapted from familiar canine neurologic examination procedures, this should confer a significant benefit. This approach governs the methodology employed for evaluation.

The following set of observational procedures is to be performed as a prestudy evaluation on all dogs (control and test). All dogs should also be screened prestudy for clinical chemistry and clinical pathology parameters, and only animals within the normal range are to be utilized in the study. The method is appropriate for all dogs on study except neonates. All data should be recorded on a score sheet. This battery is incorporated as part of a standard 30-day (or 28-day) study design, with evaluations to be performed predose and 1, 3 to 4, and 24 hours postdose on the first and last days of dosing.

Neurologic assessment consists of the following:

1. General Observations: The following assessments are to be made with the dog walking about freely in the procedure room. All end points (other than body temperature, heart rate, and where otherwise noted) are scored on a scale of 1 (least response) to 5 (greatest response), with a score of 3 denoting a normal response.
   1.1 The animal's core (rectal) body temperature
   1.2 Mental status, general behavior

      1.3 Posture and gait

      1.4 Heart rate (using stethoscope)

2. Palpatation: The head, trunk, and limbs are palpated to assess the following:

      2.1 Musculoskeletal: Head, neck, limbs, and trunk are palpated, noting any evidence of muscular atrophy (local or generalized), twitching, or other involuntary muscular actions.

3. Cranial Nerves: Routine assessment of the cranial nerves is performed as follows:

      3.1 Olfactory: Tape two conical tubes, one empty, the other with wet food inside; put one in either hand of a researcher, and then loosen caps to allow air flow and observe the animal's response to the hand with the tube containing the wet food.

      3.2 Optic/Oculomotor/Trochlear/Abducent: These nerves should be assessed by the following methods:

            3.2.1 Observe position and opening of pupil as well as the size of the pupil.

            3.2.2 Observe papillary response to light from a penlight. Observe normal physiologic nystagmus associated with lateral and dorsoventral movement of the eye.

      3.3 Trigeminal:

            3.3.1 Motor: Assess jaw tone.

            3.3.2 Sensory: Observe blink reflex after gently touching medal and lateral canthus of each eye. Observe movement of facial muscles after gently probing each nares; this also allows assessment of facial nerve motor pathways.

      3.4 Vestibulocochlear:

            3.4.1 Cochlear: Sharply clap hands with animal turned away (avoiding visual stimulation). Observe movement of ears or head in response (Preyer's reflex).

            3.4.2 Vestibular: Gait and posture previously assessed should be considered in assessment of vestibular function.

            3.4.3 Note head position.

            3.4.4 Observe eyes for abnormal nystagmus. Animal may be positioned in lateral or dorsal recumbency to assess positional nystagmus if required by protocol. Each eye should be observed for normal physiologic nystagmus following lateral or dorsal–ventral movements of the head.

            3.4.5 Glossopharyngeal/Vagus: Carefully elicit a gag reflex using a finger or tongue depressor in the pharyngeal region.

            3.4.6 Hypoglossal: Normal position of tongue should be observed after wetting the animal's nose with a water-soaked cotton ball. If the animal fails to lick in response to wetting the nose, tongue position can be examined by opening the animal's mouth.

4. Postural Reactions: The following assessments are to be performed on an examination table, procedural table, or floor with no-slip pads or mats placed to provide traction for the animal.

4.1 Proprioceptive Positioning: The animal is supported by the examiner or the assistant. Each paw is gently flexed and placed so its cranial surface touches the examination table, supporting the animal's weight with the other hand.

4.2 Wheelbarrow: Hind limbs are supported by the examiner as the animal is moved forward, allowing the animal to walk using its forelimbs.

4.3 Hemistanding: Each set of opposing (front and back) limbs are supported by the examiner as the animal is allowed to stand on the remaining limbs.

4.4 Hopping: Each set of ipsilateral limbs are supported by the examiner as the animal is moved away from the elevated limbs, causing the animal to hop laterally on the other limbs. Examiner should take care to note movement of each standing limb during procedure.

4.5 Hemiwalking: Each set of opposing limbs are supported by the examiner as the animal is moved forward, causing the animal to walk on the other limbs. Examiner should take care to note movement of each limb during procedure.

4.6 Extensor Postural Thrust: The animal is supported around the thorax or with the examiner's hand under each auxillary area and lowered until the hind paws contact the traction mats.

4.7 Tactile Placing: While the animal's eyes are covered, the animal is supported by its chest and moved slowly toward the edge of the table, allowing the cranial aspect of the distal limb (metacarpal region) to contact the edge. Placement of the front limbs up onto the examination table should be observed.

4.8 Placing, Visual: While the animal's eyes are not covered, the animal is supported by its chest, and moved slowly toward the edge of the table, allowing the cranial aspect of the distal limb (metacarpal region) to contact the edge. Placement of each of the front limbs up onto the table should be observed (the animal may place its front limbs onto the table before actually touching it).

4.9 Tonic Neck: With animal in standing position, elevate the animal's head and extend the neck, observing front and hind limbs.

5. Spinal Reflexes:

5.1 Myotatic: Myotatic reflexes are performed with the animal in lateral recumbency.

5.1.1 Quadriceps: Rear leg is supported by examiner, and stifle is slightly flexed. Straight patellar ligament is sharply struck with plexor, and the limb is observed for extension of the stifle. This will normally be the only myotic reflex performed.

5.1.2 Cranial Tibial / Gastrocnemius / Extensor Carpi Radialis / Triceps Brachii: Not normally assessed unless lesions are suspected or are required by protocol. Examiner should be experienced in producing these particular reflexes, as proper interpretation can be difficult.

5.1.3 Flexor (withdrawal): With the animal in lateral recumbency, a toe on each foot is gently pinched using a forceps, and each corresponding limb is observed for flexion (away from stimulus). Minimal noxious stimulus to produce withdrawal should be used.

5.1.4 Extensor Thrust: With the animal in lateral recumbency, each limb is supported by the examiner in slight flexion. The toes are gently spread, and slight pressure is applied with a finger in between the pads.

5.1.5 Crossed Extensor: With the animal in lateral recumbency, each limb is sharply flexed by proximal movement. The contralateral limb is observed for an extension reflex.

5.1.6 Extensor Toe: Not normally assessed unless specific lesions are suspected or protocol indicated.

5.1.7 Perineal: The perineum is gently pinched using hemostat, while the examiner observes anal sphincter contraction and/or tail movement.

6. Sensation:

6.1 Deep Pain: Toes are pinched using forceps. Care should be taken to prevent bruising or tissue damage while eliciting this response. The observer looks for any behavioral response to this stimulus, except withdrawal of the limb (reflex).

6.2 Touch Response: With an assistant holding the front of the dog in a standing position, the response to a slight touch to the body out of sight of the dog is noted.

7. Limb Strength:

7.1 Fore and hind limb strengths are evaluated by grasping the appropriate limb, elevating this limb to the animal's neck level, then "wheelbarrowing" it forward for two body lengths (~6 inches). Ability to perform this task is scored as normal (3), reduced (2), and impaired (1) (frequently stumbles), and absent (0) (animal cannot support weight or perform task).

5.4 Sleep induction and interaction with hypnotics

Sleep induction in rodents can be evaluated by loss of the righting reflex. Intrinsic sleep-inducing activity can be observed during the Irwin test or FOB (see above). A more sensitive index of sleep enhancement can, however, be obtained by observing the interaction of the test substance with sleep induced by standard hypnotics such as barbiturates. A further advantage of this kind of procedure is that it can also reveal antisleep activity (decreased duration or abolition of barbiturate-induced sleep). These studies therefore provide a useful complement to studies of general behavior and spontaneous locomotion. Drug interaction procedures have the disadvantage that effects observed could arise through pharmacokinetic factors such as changes in metabolism, rather than through a true pharmacodynamic interaction. For this reason, interaction tests using barbital as standard barbiturate are to be recommended because barbital itself undergoes no metabolism, in contrast to most other barbiturates.

5.5 Higher cognitive function

More complex tests of higher cognitive function find their place during the latter phases of CNS safety assessment because they are more time-consuming to perform and cannot therefore be performed on a routine screening basis. Included under the term *cognition* are learning, memory and attention, and general intellectual activity. It is important that drugs should be devoid of impairing effects on these functions, whatever their indication. Indeed package inserts for many kinds of drugs contain warnings about the use of the drug while driving, working, or engaging in various daily occupations. It is therefore essential that CNS safety pharmacology provides procedures for evaluating these effects in animal studies. There is no standardized set of procedures in this area. The following provides an indication of the kinds of experiment that could be undertaken. The list is by no means exhaustive.

5.5.1 Passive avoidance

One of the simplest procedures for looking for adverse effects on learning and memory is the so-called one-trial passive avoidance task [15]. There are several variants of this procedure, but the basic principal is that a rat or a mouse receives an aversive stimulation in the recognizable environment and on a later occasion shows it has remembered by not going there (passive avoidance). For example, a rat placed into the lit compartment of a two-compartment box will explore the apparatus and eventually enter the dark compartment where it receives a brief electric shock [19]. When placed again into the lit compartment 24 or 48 h later, the rat will avoid going into the dark compartment despite a natural preference for the dark. Amnesia-inducing drugs (e.g., benzodiazepines, anticholinergics, NMDA antagonists) administered before the first exposure will attenuate the animal's memory for the shock as shown by a decreased latency to enter the dark compartment on the test day.

Passive avoidance is a fairly simple and rapid procedure because it involves learning obtained in a single trial. It is therefore suitable as a first screen for potential cognition impairing activity. On the other hand, it is not very clearly interpretable in terms of the cognitive function implicated as it is not possible to distinguish a drug's effects on learning or memory, or whether the drug influenced performance by impairing attention or even pain sensitivity during the learning trial.

5.5.2   *Morris maze*

Another fairly simple procedure, which allows a greater degree of interpretation, is the Morris water maze. The water maze task has been most extensively used to investigate specific aspects of spatial memory. This task is based upon the premise that animals have evolved an optimal strategy to explore their environment and escape from the water with a minimum amount of effort—i.e., swimming the shortest distance possible. The time it takes a rat to find a hidden platform in a water pool after previous exposure to the setup, using only available external cues, is determined as a measure of spatial memory. Studies with pharmacologic agents are initiated when performance is stable, and the water maze task is particularly sensitive to the effects of aging [20]. Alternatively, one can study the effects of a drugs or lesions upon the acquisition of this task. Drugs can be administered (or lesions produced) prior to training to assess their effects upon acquisition, or after acquisition to assess their effects upon performance. A water maze can easily be built or purchased by an investigator, so the cost for this equipment can be quite low. Using the maze can be labor intensive, requiring that a tester be present or nearby throughout the task. This maze is typically used for rats, but it can be scaled down in size for use with mice.

5.5.2.1   *Materials*
Rats
Pharmacologic or toxicologic agents (optional)
Water maze apparatus
Tracking system and software (such as provided by Columbus Instruments, HVS Image, San Diego Instruments, or CPL Systems)

5.5.2.2   *Set up apparatus and begin acquisition testing*

1. Set up water maze. A water-tight pool, painted white, should be positioned in a room with various external cues that are visible to a rat swimming in the pool, such as a doorway, overhead lights and camera (if desired), and large simple designs on the walls. Make water opaque by adding powdered milk or nontoxic white paint to the water. The pool should be designed so that it can be easily drained on a regular basis.
2. Insert platform into one quadrant of the pool.
3. Place rat into water with its head pointed toward the side of pool. The starting position should be at a different, and randomized, location each day of testing—north, south, and so forth.
4. Record time (in seconds) it takes the rat to find the submerged platform. Guide rat to platform on first few trials if it requires > 120 sec.

    A tracking camera, positioned ~ 200 cm above the center of the pool can be used to quantify the distance swam on each trial and thereby determine swimming speed when combined with latency measurements. The tracking system can also display swim path and distance and proved additional information on search efficiency and exploration patterns during acquisition and probe trials. This equipment and associated computer software can be obtained from several commercial manufacturers.
5. Allow rat to remain on platform for 10 to 15 sec. This allows the rat time to return to an appropriate place at the side of the water pool to be ready for Step 6.
6. Remove rat from pool. Wait 5 min.
7. Release rat into pool (from same location) with platform in same location. Record time for rat to find platform.
8. Give each rat 4 trials on the first day.

5.5.2.3   *Continue trials*

1. On second day, insert platform in same location as on the first day.
2. Release rat with its head pointed toward the side of the water pool.
3. Record time it takes rat to find platform.
4. Give rat 8 to 10 trials per day with 5-min intertribal intervals for several days until performance is stable and latency to find the platform is low (<5 to 7 sec).
5. Perform data analysis. Performance is expressed as the average time it takes each rat to find the submerged platform. The data are best presented as a line drawing comparing the latency to find the platform for each group versus daily test sessions. Data from 2 or 4 days of testing can be averaged into blocks.
6. Repeat sequence with test and control standard drug.

5.6   Isolated tissue assays

The classic approach to screening for nervous system effect is a series of isolated tissue preparation bioassays, conducted with appropriate standards, to determine if the material acts pharmacologically directly on neural receptor sites or transmission properties. Though these bioassays are normally performed by a classical pharmacologist, a good technician can be trained to conduct them. The required equipment consists of a Mangus (or similar style) tissue bath [18,21,22] a physiograph or kymograph, force transducer, glassware, a stimulator, and bench spectrophotometer. The assays used in the screening battery are listed in Table 5.5, along with the original reference describing each preparation and assay. The assays are performed as per the original author's descriptions with only minor modifications, except that control standards (as listed in Table 5.5) are always used. Only those assays that are appropriate for the neurological/muscular alterations observed in the screen are used. Note that all these are intact organ preparations, not minced tissue preparations as others [17] have recommended for biochemical assays.

The first modification in each assay is that, where available, both positive and negative standard controls (pharmacological agonists and antagonists, respectively) are employed. Before the preparation is used to assay the test material, the tissue preparation is exposed to the agonist to ensure that the preparation is functional and to provide a baseline dose response curve against which the activity of the test material can be quantitatively compared. After the test material has been assayed (if a dose response curve has been generated), one can determine whether the antagonist will selectively block the activity of the test material. If so, specific activity at that receptor can be considered as established. In this assay sequence, it must be kept in mind that a test material may act to either stimulate or depress activity, and therefore the roles of the standard agonists and antagonists may be reversed.

Commonly overlooked when performing these assays is the possibility of metabolism to an active form that can be assessed in this *in vitro* model. The test material should be tested in both original and *metabolized* forms. The metabolized form is prepared by incubating a 5% solution (in aerated Tyrodes) or other appropriate physiological salt solution with strips of suitably prepared test species liver for 30 min. A filtered supernatant is then collected from this incubation and tested for activity. Suitable metabolic blanks should also be tested. This is a classic nervous system pharmacology approach.

5.6.1   Electrophysiology methods

A number of electrophysiological techniques available can be used to detect and/or assess neurotoxicity. These techniques can be divided into

Table 5.5 Isolated Tissue Pharmacologic Assays

Assay system	End point	Standards (agonist/antagonist)	References
Rat Ileum	General activity	None (side-spectrum assay for intrinsic activity)	15
Guinea Pig *Vas Deferens*	Muscarinic nicotinic or muscarinic	Methacholine/atropine Methacholine/hexamethonium Methacholine/atropine	22
Rat Serosal Strip	Nicotinic	Methacholine/hexamethonium	23
Rat *Vas Deferens*	Alpha adrenergic	Norepinephrine/phenoxybenzamine	24
Rat Uterus	Beta adrenergic	Epinephrine/propanol	25
Rat Uterus	Kinin receptors	Bradykinin/none	26
Guinea Pig Tracheal Chain	Dopaminergic	Dopamine/none	15
Rat Serosal Strips	Tryptaminergic	5-Hydroxytryptamine (serotonin)/dibenzyline or lysergic acid dibromide	27
Guinea Pig Tracheal Chain	Histaminergic	Histamine/benadryl	28, 29
Guinea Pig Ileum (electrically stimulated)	Endorphin receptors	Methenkephaline/none	30
Red Blood Cell Hemolysis	Membrane stabilization	Chloropromazine (not a receptor-mediated activity)	31
Frog Rectus Abdominis	Membrane depolarization	Decamethonium iodide (not a receptor-mediated activity)	32

two broad general categories: those focused on CNS function and those focused on peripheral nervous system function [34].

First, however, the function of the individual components of the nervous system, how they are connected together, and how they operate as a complete system should be very briefly overviewed.

Data collection and communication in the nervous system occurs by means of graded potentials, action potentials, and synaptic coupling of neurons. These electrical potentials may be recorded and analyzed at two different levels depending on the electrical coupling arrangements: individual cells (that is, intracellular and extracellular) or multiple cell (e.g., EEG, evoked potentials (EPs), slow potentials). These potentials may be recorded in specific central or peripheral nervous system areas (e.g., visual cortex, hippocampus, sensory and motor nerves, muscle spindles) during various behavioral states or in *in vitro* preparations (e.g., nerve-muscle, retinal photoreceptor, brain slice).

5.7　CNS function: Electroencephalography

The EEG is a dynamic measure reflecting the instantaneous integrated synaptic activity of the CNS, which most probably represents, in coded form, all ongoing processes under higher nervous control. Changes in frequency, amplitude, variability, and pattern of the EEG are thought to be directly related to underlying biochemical changes, which are believed to be directly related to defined aspects of behavior. Therefore, changes in the EEG should be reflected by alterations in behavior and vice versa.

The human EEG is easily recorded and readily quantified, is obtained noninvasively (scalp recording), samples several regions of the brain simultaneously, requires minimal cooperation from the subject, and is minimally influenced by prior testing. Therefore, it is a very useful and recommended clinical test in cases in which exposure to drugs produces symptoms of CNS involvement and in which long-term exposures to high concentrations are suspected of causing CNS damage.

Since the EEG recorded using scalp electrodes is an average of the multiple activity of many small areas of cortical surface beneath the electrodes, it is possible that in situations involving noncortical lesions, the EEG may not accurately reflect the organic brain damage present. Noncortical lesions following acute or long-term low-level exposures to toxicants are well documented in neurotoxicology [15]. The drawback mentioned earlier can be partially overcome by utilizing activation or evocative techniques, such as hyperventilation, photic stimulation, or sleep, which can increase the amount of information gleaned from a standard EEG.

As a research tool, the utility of the EEG lies in the fact that it reflects instantaneous changes in the state of the CNS. The pattern can thus be used to monitor the sleep-wakefulness cycle activation or deactivation of the brainstem, and the state of anesthesia during an acute electrophysiological procedure. Another advantage of the EEG, which is shared by all CNS electrophysiological techniques, is that it can assess the differential effects of toxicants (or drugs) on various brain areas or structures. Finally, specific CNS regions (e.g., the hippocampus) have particular patterns of after-discharge following chemical or electrical stimulation, which can be quantitatively examined and utilized as a tool in neurotoxicology. It also serves well in primates to determine if a drug has penetrated and "hit" target receptors in the brain.

The EEG does have some disadvantages, or, more correctly, some limitations. It cannot provide information about the effects of toxicants on the integrity of sensory receptors or of sensory or motor pathways. As a corollary, it cannot provide an assessment of the effects of toxicants on sensory system capacities. Finally, the EEG does not provide specific information at the cellular level and therefore lacks the rigor to provide detailed mechanisms of action.

Rats represent an excellent model for this as they are inexpensive, resist infection during chronic electrode and cannulae implantation, and are relatively easy to train so that behavioral assessments can be made concurrently.

Depending on the time of drug exposure, the type of scientific information desired, and the necessity of behavioral correlations, a researcher can perform acute and/or chronic EEG experiments. Limitations of the former are that most drugs that produce general anesthesia modify the pattern of EEG activity and thus can complicate subtle effects of toxicants. However, this limitation can be partially avoided if the effect is robust enough. For sleep-wakefulness studies, it is also essential to monitor and record the electromyogram (EMG).

Excellent reviews of these electrophysiology approaches can be found in Fox et al. [35] and Takeuchi and Koike [36].

5.8   Neurochemical and biochemical assays

Though some very elegant methods are now available to study the biochemistry of the brain and nervous system, none has yet discovered any generalized marker chemicals that will serve as reliable indicators or early warnings of neurotoxic actions or potential actions. There are, however, some useful methods. Before looking at these, however, one should understand the basic problems involved.

Normal biochemical events surrounding the maintenance and functions of the nervous system center around energy metabolism; biosynthesis of macromolecules; and neurotransmitter synthesis, storage, release, uptake, and degradation. Measurement of these events is complicated by the sequenced nature of the components of the nervous system and the transient and labile nature of the moieties involved. Use of measurements of alternations in the these functions as indicators of neurotoxicity is further complicated by our lack of a complete understanding of the normal operation of these systems and by the multitude of day-to-day occurrences (such as diurnal cycle, diet, temperature, age, sex, and endocrine status) that are constantly modulating the baseline system. For detailed discussions of these difficulties, the reader is advised to see Damstra and Bondy [37,38].

References

1. CPMP, *Note for Guidance on Safety Pharmacology Studies in Medicinal Product Development*, 1998.
2. ICH, *Safety Pharmacology Studies for Human Pharmaceuticals*, 2002.
3. Irwin, S., Comprehensive observational assessment: la. A systematic, quantitative procedure for assessing the behavioral and physiologic state of the mouse, *Psychopharmacologia*, Berlin, 1968, 13:222–257.
4. Gad, S.C., A neuromuscular screen for use in industrial toxicology, *J. Toxicol. Environ. Health*, 1982, 9:691–704.
5. Haggerty, G.C., Strategies for and experience with neurotoxicity testing of new pharmaceuticals, *J. Am. Coll. Toxicol.*, 1991, 10:677–687.
6. Mattson, J.L., Spencer, P.J., and Albee, R.R., A performance standard for clinical and functional observational battery examination of rats, *J. Am. Coll. Toxicol.*, 1996, 15:239.
7. Moser, V.C., Neurobehavioral screening in rodents (Unit 11.2). In: *Current Protocols in Toxicology*, (Maines, M.D., et al., eds.), New York: Wiley, 1999, pp. 11.2.1–11.2.16.
8. Moscardo, E., et al., An optimized methodology for the neurobehavioral assessment in rodents, *Journal of Pharmacological and Toxicological Methods*, 2007, 56:239–255.
9. Redfern, W.S., et al., Spectrum of effects detected in the rat functional observational battery following oral administration of non-CNS targeted compounds, *Journal of Pharmacological and Toxicological Methods*, 2005, 52:77–82.
10. Porsolt, R.D., et al., New perspectives in CNS safety pharmacology, *Fund. Clin. Pharm.*, 2002, 16:197–207.
11. Gad, S.C., Gad, S.D. and Gad, S.E., The dog functional observational battery (FOB) for use in pharmaceutical safety evaluation studies, *International Journal of Toxicology*, 2003, 22:1–8.
12. Gauvin, D.V., and Baird, T.J., A functional observational battery in non-human primates for regulatory-required neurobehavioral assessment, *Journal of Pharmaceutical and Toxicological Methods*, 2008, 58:88–93.
13. Zhu, Y., et al., Analysis of neurobehavioral screening date: dose-time-response modeling of continuous outcomes. *Regulatory Toxicology and Pharmacology* 2005, 41:240–255.
14. MHW, *Guidelines for Toxicity Studies of Drugs*, 1999.
15. Norton, S., Toxic responses of the central nervous system. In: *Toxicology: The Basic Science of Poisons*, 2nd ed. (Doull, J., Klaassen, C.D., and Amdur, M.O., eds.), New York: Macmillan, 1980.
16. Domer, F.R., *Animal Experiments in Pharmacological Analysis*, Springfield, IL: Charles C. Thomas, 1971, pp. 98, 115, 155, 164, 220.
17. Bondy, S.C., Rapid screening of neurotoxic agents by *in vivo* means. In *Effects of Food and Drugs on the Development and Function of the Nervous System: Methods for Predicting Toxicity* (Gryder, R.M., and Frankos, V.H., eds.), Washington, D.C.: Office of Health Affairs, FDA, 1979, pp. 133–143.
18. Turner, R.A., *Screening Methods in Pharmacology*, vols. I and II, New York: Academic, 1965, pp. 42–47, 60–68, 27–128.
19. Brady, J.V., and Lukas, S.E., *Testing Drugs for Physical Dependence Potential and Abuse Liability*. NIDA Research Monograph 52, 1984.

20. Brandeis, R., Brandys, Y., and Yehuda, S., The use of the Morris water maze in the study of memory and learning, *Int. J. Neuroscience*, 1989, 48:29–69.
21. Offermeier, J., Ariens, E.J., and Serotonin, I., Receptors involved in its action, *Arch. Int. Pharmacodyn. Ther.*, 1966, 64:92–215.
22. Nodine, J.H., and Seiger, P.E., *Animal and Clinical Pharmacologic Techniques in Drug Evaluation*, Chicago: Year Book Medical Publishers, Inc., 1964, pp. 36–38.
23. Leach, G.D.H., Estimation of drug antagonisms in the isolated guinea pig vas deferens, *J. Pharm. Pharmacol*, 1956, 8:501.
24. Khayyal, M.T., et al., A sensitive method for the bioassay of acetylcholine, *Eur. J. Pharmacol.*, 1974, 25:287–290.
25. Rossum, J.M. van, Different types of sympathomimetic β–receptors, *J. Pharm. Pharmacol.*, 1965, 17:202.
26. Levy, B., and Tozzi, S., The adrenergic receptive mechanism of the rat uterus, *J. Pharmacol. Exp. Ther.*, 1963, 142:178.
27. Gecse, A., Zsilinksky, E., and Szekeres, L., Bradykinin antagonism. In *Kinins: Pharmacodynamics and Biological Roles* (Sicuteri, R., Back, N., and Haberland, G., eds.), New York: Plenum Press, 1976, pp. 5–13.
28. Lin, R.C.Y., and Yeoh, T.S., An improvement of Vane's stomach strip preparation for the assay of 5-hydroxy-tryptamine, *J. Pharm. Pharmacol*, 1965,17: 524–525.
29. Castillo, J.C., and De Beer, E.J., The guinea pig tracheal chain as an assay for histamine agonists. *Fed. Proc.*, 1947, 6:315.
30. Castillo, J.C., and De Beer, E.J., The tracheal chain, *J. Pharmacol. Exp. Ther.*, 1947b, 90:104.
31. Cox, B.M., et al., A peptide-like substance from pituitary that acts like morphine 2, Purification and properties, *Life Sci.*, 1975, 16:1777–1782.
32. Seeman and Weinstein, *Drug Safety*, Seppalaxnier, 1975, 1966.
33. Burns, B.D., and Paton, W.D.M., Depolarization of the motor end-plate by decamethonium and acetylcholine. *J. Physiol*, London, 1951, 115:41–73.
34. Seppalaninen, A.M., Applications of neurophysiological methods in coocupational medicine: a review. *Scan. J. Work Envirn. Health*, 1975, 1:1–14.
35. Fox, D.A., Lowndes, H.E., and Bierkamper, G.G., Electrophysiological techniques in neurotoxicology. In: *Nervous System Toxicology* (Mitchell, C.L., ed.), New York: Raven Press, 1982, pp. 299–336.
36. Takeuchi, Y., and Koike, Y., Electrophysiological methods for the *in vivo* assessment of neurotoxicology. In: *Neurotoxicology* (Blum, K., and Manzo, L., eds.), New York: Marcel Dekker, Inc., 1985, pp. 613–629.
37. Damstra, T., and Bondy, S.C., The current status and future of biochemical assays for neurotoxicity. In: *Experimental and Clinical Neurotoxicology* (Spencer, P.S., and Shaumburg, H.H., eds.), Baltimore: Williams and Wilkins, 1980, pp. 820–833.
38. Damstra, T., and Bondy, S.C., Neurochemical approaches to the detection of neurotoxicity. In: *Nervous System Toxicology* (Mitchell, C.L., ed.), New York: Raven Press, 1982, pp. 349–373.

chapter six

Respiratory system

As early as 1964, it became apparent that β-adrenergic blocking agents could lead to bronchoconstriction (and possible death) in asthmatics [1]. Since then, many similar adverse effects have been identified. These known effects of drugs from a variety of pharmacologic/therapeutic classes on the respiratory system are summarized in Tables 6.1, 6.2, and 6.3. Resulting worldwide regulatory requirements (Tables 6.4, 6.5) support the need for conducting respiratory/pulmonary evaluations in safety pharmacology [16]. The objective of such studies is to evaluate the potential for drugs to cause secondary pharmacologic or toxicologic effects that influence respiratory function. Changes in respiratory function can result either from alterations in the pumping apparatus that controls the pattern of pulmonary ventilation or from changes in the mechanical properties of the lung that determine the transpulmonary pressures (work) required for lung inflation and deflation.

The respiratory system is responsible for generating and regulating the transpulmonary pressures needed to inflate and deflate the lung. Normal gas exchange between the lung and blood requires breathing patterns that ensure appropriate alveolar ventilation. Ventilatory disorders that alter alveolar ventilation are defined as hypoventilation or hyperventilation syndromes. Hyperventilation results in an increase in the partial pressure of arterial CO_2 above normal limits and can lead to acidosis, pulmonary hypertension, congestive heart failure, headache, and disturbed sleep. Hypoventilation results in a decrease in the partial pressure of arterial CO_2 below normal limits and can lead to alkalosis, syncope, epileptic attacks, reduced cardiac output, and muscle weakness.

Normal ventilation requires that the pumping apparatus provide both adequate total pulmonary ventilation (minute volume) and the appropriate depth (tidal volume) and frequency of breathing. The depth and frequency of breathing required for alveolar ventilation are determined primarily by the anatomic deadspace of the lung. In general, a rapid shallow breathing pattern (tachypnea) is less efficient than a slower deeper breathing pattern that achieves the same minute volume. Thus, any change in minute volume, tidal volume, or the rate of breathing can influence the efficiency of ventilation [17]. The inspiratory and expiratory phases of individual breath rates of airflow and durations are distinct and independently controlled [18]. Thus, by characterizing changes in the

Table 6.1 Agents Known to Cause Pulmonary Disease [2–8]

Chemotherapeutic	Analgesics
Cytotoxic	Heroin*
Azathioprine	Methadone*
Bleomycin*	Noloxone*
Busulfan	Ethchlorvynol*
Chlorambucsil	Propoxyphene*
Cyclophosphamide	Salicylates*
Etoposide	Cardiovascular
Melphalan	Amiodarone*
Mitomycin*	Angiotensin-converting enzyme
Nitrosoureas	inhibitors
Procarbazine	Anticoagulants
Vinblastine	Beta blockers*
Ifosfamide	Dipyridamole
Noncytotoxic	Fibrinolytic agents*
Methotrexate*	Protamine*
Cytosine arabinoside*	Tocainide
Bleomycin*	Inhalants
Procarbazine*	Aspirated oil
Antibiotic	Oxygen*
Amphotericin B*	Intravenous
Nitrofurantoin	Blood*
Acute*	Ethanolamine oleate (sodium
Chronic	morrhuate)*
Sulfasalazine	Ethiodized oil (lymphangiogram)
Sulfonamides	Talc
Pentamidine	Fat emulsion
Anti-inflammatory	Miscellaneous
Acetylsalicylic acid*	Bromocripitine
Gold	Dantrolene
Methotrexate	Hydrochlorothiazide*
Nonsteroidal anti-inflammatory	Methysergide
agents	Oral contraceptives
Penicillamine*	Tocolytic agents*
Immunosuppressive	Tricyclics*
Cyclosporine	L-Tryptophan
Interleukin-2*	Radiation
	Systemic lupus erythematosus (drug
	induced)*
	Complement-mediated leukostasis*

* Typically cause acute or subacute respiratory insufficiency.

airflow rate and duration of each of these phases, mechanisms responsible for changes in tidal volume or respiratory rate can be identified [19,20]. For example, a decrease in airflow during inspiration (the active phase) is generally indicative of a decrease in respiratory drive, while a decrease in

Table 6.2 Drugs That Adversely Affect Respiratory Function [1–3,9–14]

Drugs known to cause or aggravate bronchospasm	Agents associated with pleural effusion
Vinblastine	Chemotherapeutic agents
Nitrofurantoin (acute)	Nitrofurantoin (acute)
Acetylsalicylic acid	Bromocriptine
Nonsteroidal anti-inflammatory agents	Dantrolene
Interleukin-2	Methysergide
Beta-blockers	L-Tryptophan
Dipyridamole	Drug-inducing systemic lupus
Protamine	erythematosus
Nebulized pentamidine, beclomethasone, and propellants	Tocolytics
Hydrocortisone	Amiodarone
Cocaine	Esophageal variceal sclerotherapy agents
Propafenone	Interleukin-2

Agents associated with acute-onset pulmonary insufficiency*	Agents that cause subacute respiratory failure
Bleomycin plus O2	Chemotherapeutic agents
Mitocycin	Nitrofurantoin (chronic)
Bleomycin†	Amiodarone
Procarbazine†	L-Trytophan
Methotrexate†	Drug-inducing systemic lupus erythematosus
Amphotericin B	
Nitrofurantoin (acute)‡	
Acetylsalicylic acid‡	
Interleukin-2‡	
Heroin and other narcotics‡	
Epinephrine‡	
Ethchlorvynol‡	
Fibrinolytic agents	
Protamine	
Blood products ‡	
Fat emulsion	
Hydrochlorothiazide	
Complement-mediated leukostasis	
Hyskon (dextran-70) ‡	
Tumor necrosis factor‡	
Intrathecal methotrexate	
Tricyclic antidepressants‡	
Amiodarone plus 02	
Naloxone	

* Onset at less than 48 h.
† Associated with hypersensitivity with eosinophilia.
‡ Usually reversible within 48–72 h, implying noncardiac pulmonary edema rather than inflammatory interstitial pneumonitis.

Table 6.3 Drugs Known to Influence Ventilatory Control [15]

Depressants	Stimulants
Inhaled anesthetics	Alkaloids
Barbiturates	Nicotine
Benzodiazepines	Lobeline
Diazepam	Piperdine
Temazepam	Xanthine analogs
Chlordiazepoxide	Theophylline
Serotonin analogs	Caffeine
Methoxy-(dimethyl)-tryptamine	Theobromine
Dopamine analogs	Analeptics
Apomorphine	Doxapram
Adenosine analogs	Salicylates
2-Chloroadenosine	Progesterone analogs
R-Phenylisopropyl-adenosine (R-PIA)	Almitrine
N-Ethylcarboxamide (NECA)	Glycine analogs
B-Adrenergic antagonists	Strychnine
Timolol maleate	GABA antagonists
GABA analogs	Picrotoxin
Muscimol	Bicuculline
Baclofen	Serotonin synthesis
Opiates	inhibitors
Morphine	p-Chlorophenylalanine
Codeine	Reserpine
Methadone	
Meperidine	
Phenazocine	
Tranquilizers/analgesics	
Chlorpromazine	
Hydroxyzine	
Rompum (xylazine)	
Nalorphine	

Table 6.4 Required Respiratory System Safety Pharmacology Evaluation

Respiratory functions

Measurement of rate and relative tidal volume in conscious animals

Pulmonary function

Measurement of rate, tidal volume, and lung resistance and compliance in
  anesthetized animals

Table 6.5 Regulatory Documents Recommending Respiratory Function
Testing in Safety Drug Studies

United States	*FDA Guideline for the Format and Content of the Nonclinical Pharmacology/Toxicology Section of an Application,* Section IID, p. 12, Feb. 1987.
Japan	*Ministry of Health and Welfare Guidelines for Safety Pharmacology Studies Required for the Application for Approval to Manufacture (Import) Drugs,* Notification YAKUSHIN-YAKU No. 4, Jan. 1991.
Australia	*Guidelines for Preparation and Presentation of Applications for Investigational Drugs and Drug Products Under the Clinical Trials Exemption Scheme,* pp. 12, 15, 2002.
Canada	RA5 Exhibit 2, *Guidelines for Preparing and Filing Drug Submissions,* p. 21, 1987.
United Kingdom	Medicines Act 1968, *Guidance Notes on Applications for Product Licenses,* MAL 2, p. A3F-1.
ICH	*S7A Safety Pharmacology Assessment of New Human Pharmaceuticals,* June 2001.

Results [2,3]

Parameters	Theophylline 10 mg/kg PO	Pentobarbital 35 mg/kg IP	Diazepam 35 mg/kg IP	Codeine 100 mg/kg IP
F(beat/min)	+ + +	− − −	− − −	No Change
TV (ml)	No Change	No Change	No Change	−
Ti (s)	− −	+ +	+ +	+
Te (s)	− −	+ + +	+ +	−
PIF (ml/s)	+ +	−	−	−
PEF (ml/s)	+ +	No Change	+	−
Penh	−	+	+	No Change

Note: Qualitative increase (+) or decrease (−) in value.

airflow during expiration (the passive phase) is generally indicative of an obstructive disorder.

Mechanisms of ventilatory disorders can also be characterized as either central or peripheral. Central mechanisms involve the neurologic components of the pumping apparatus that are located in the central nervous system and include the medullary central pattern generator (CPG) as well as integration centers located in the medulla, pons, hypothalamus, and cortex of the brain that regulate the output of the CPG [18]. The major neurologic inputs from the peripheral nervous system that influence the CPG are the arterial chemoreceptors [18]. Many drugs stimulate or depress ventilation by selective interaction with the central nervous system [9–11,21,22] or arterial chemoreceptors [23,24].

Defects in the pumping apparatus are classified as hypo- or hyperventilation syndromes and are best evaluated by examining ventilatory parameters in a conscious animal model. The ventilatory parameters include respiratory rate, tidal volume, minute volume, peak (or mean) inspiratory flow, peak (or mean) expiratory flow, and fractional inspiratory time. Defects in mechanical properties of the lung are classified as obstructive or restrictive disorders and can be evaluated in animal models by performing flow-volume and pressure-volume maneuvers, respectively. The parameters used to detect airway obstruction include peak expiratory flow, forced expiratory flow at 25% and 75% of forced vital capacity, and a timed forced expiratory volume; while the parameters used to detect lung restriction include total lung capacity, inspiratory capacity, functional residual capacity, and compliance. Measurement of dynamic lung resistance and compliance, obtained continuously during tidal breathing, is an alternative method for evaluating obstructive and restrictive disorders, respectively, and is used when the response to drug treatment is expected to be immediate (within minutes post-dose). The species used in the safety pharmacology studies are the same as those generally used in toxicology studies (rats and dogs) since pharmacokinetic and toxicologic/pathologic data are available in these species. These data can be used to help select test measurement intervals and doses and to aid in the interpretation of functional change. The techniques and procedures for measuring respiratory function parameters are well-established in guinea pigs, rats, and dogs [21,25–29].

The key questions in safety pharmacology of the respiratory system are as follows:

- Does the substance affect the mechanisms of respiratory control (central or peripheral) leading to hypoventilation (respiratory depression) or hyperventilation (respiratory stimulation)?
- Does the substance act on a component of the respiratory system to induce, for example, bronchospasm, obstruction, or fibrosis?
- Does the substance induce acute effects or we can expect chronic effects?
- Are the effects observed dose dependent or independent?

6.1   Plethysmography

The classic approach to measuring respiratory function in laboratory animals is plethysmography. It has two basic governing principles [18,22,30]:

- The animal (mouse, rat, or dog), anesthetized or not, restrained or not, is placed in a chamber (simple or double) with pneumotachographs.
- The variations of pressure in chamber(s) at the time of the inspiration and the expiry make it possible to obtain the respiratory flow of the animal.

There are three main types of body plethysmographs: constant volume, constant pressure, and pressure volume. The constant-volume body plethysmograph is a sealed box that detects volume change by measurement of pressure changes inside the box. While inside the plethysmograph, inhalation of room air (from outside the plethysmograph) by the test animal induces an increase in lung volume (chest expansion) and thus an increase in plethysmograph pressure. On the other hand, exhalation to the atmosphere (outside the plethysmograph) induces a decrease in the plethysmograph pressure. The magnitude of lung volume change can be obtained via measurement of the change in plethysmographic pressure and the appropriate calibration factor. The plethysmograph is calibrated by injecting or withdrawing the change in box pressure. To avoid an adiabatic artifact, the rate of air injection or withdrawal is kept the same as that of chest expansion, indicated by the same dP/dt (change in pressure over change in time).

The constant-pressure body plethysmograph is a box with a pneumotachograph port built into its wall. This plethysmograph detects volume change via integration of the flow rate, which is monitored by the pneumotachograph port. There is an outward flow (air moving from the plethysmograph to the atmosphere) during inspiration and inward flow during expiration. Alternatively, in place of a pneumotachograph, a spirometer can be attached to the constant pressure plethysmograph to detect volume changes. For detection of plethysmographic pressure and flow rate, sensitive pressure transducers are usually employed. It is important that the transducer be capable of responding to volume changes in a linear fashion within the volume range studied. The plethysmograph should have negligible leaks, and temperature should not change during all respiratory maneuvers. The plethysmograph should also have linear characteristics with no hysteresis. Dynamic assuracy requires an adequate frequency response. A fast integrated flow plethysmograph, with a flat amplitude response for sinusoidal inputs up to 240 Hz, has been developed for rats, mice, and guinea pigs [19, 35]. Similar plethysmographs can also be provided for use with large mammals.

A third type of pressure-volume plethysmograph has the mixed characteristics of the two types of body box mentioned above. For a constant-pressure plethysmograph, the change in volume at first is associated with gas compression or expansion. This fraction of the volume change can be corrected by electronically adding the plethysmographic pressure change to the volume signal. Therefore, the combined pressure-volume plethysmograph has excellent frequency response characteristics and a wide range of sensitivities [20].

If volume, flow rate, and pressure changes are detected at the same time, several respiratory variables can be derived simultaneously from the

TABLE 6.6 Pulmonary Variables from the
Maximal Expiratory Flow-Volume Curve

Test description	Recommended term	Symbol
1. Volume of gas expired after full inspiration expiration being as rapid and complete as possible (i.e., forced)	Forced vital capacity	FVC
2. Peak expiratory flow (liter/min or liter/s) measured by various instruments	Peak expiratory flow, qualified by name of instrument used	PEF
3. Volume of gas exhaled over a given time interval during a complete forced expiration	Forced expiratory volume, qualified by time interval in s	FEV_t
4. FEV_t expressed at % of FVC	Percent of FVC expired in time interval t	$FEV_t/FVC\%$
5. Volume of air exhaled over a specified volume range of the FVC divided by the time to exhale this volume, expressed as liter/min or liter/s. Examples:	Mean forced expiratory flow between two designated volume points in FVC	
Volume between 0.2 and 1.2 liters of the FVC/time	Mean forced expiratory flow, 0.1–1.2 1	$FEF_{0.2-1.2\ liters}$
Volume between 25% and 75% of the FVC time of FVC	Mean forced expiratory flow, 25%–75%	$FEF_{25\%-75\%}$
6. Maximal expiratory flow at a specific volume during FVC expressed in liter/min or liter/s Examples:	Maximal expiratory flow qualified at xx percent of VC (Note: 100% VC is a TLC; 0% is at RV)	
Flow at point when 75% of FVC remaining	Maximal expiratory flow, 75% of VC	$Vmax_{75}$
Flow at point when 50% of FVC remaining	Maximal expiratory flow, 50% of VC	$Vmax_{50}$
Flow at point when 25% of FVC remaining	Maximal expiratory flow, 25% of VC	$Vmax_{25}$

raw signals. The whole-body plethysmograph method can then be used to measure most respiratory variables, such as tidal volume, breathing frequency, minute ventilation, compliance, pulmonary resistance, functional residual capacity, pressure-volume characteristics, and maximal expiratory flow-volume curves. Table 6.6 defines the parameters that are typically determined by these methods, and Figure 6.1 shows how they actually appear in tracings.

Parameters

- Respiratory frequency (f)
- Inspiratory time (Ti)
- Expiratory time (Te)
- Peak inspiratory flow (PIF)
- Peak expiratory flow (PEF)
- Tidal volume (TV)
- Pulmonary resistance: R=DP/DF
- Enhanced pause (Penh):
  [Te/40% Tr-1] × PEF/PIF × 0.67

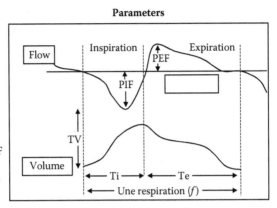

Figure 6.1 Respiratory parameters.

Selection of the proper reference values for interpretation of findings is essential [31,32].

6.2   Design of respiratory function safety studies

6.2.1   General considerations

The objective of a safety pharmacology evaluation of the respiratory system is to determine whether a drug has the potential to produce a change in respiratory function. Since a complete evaluation of respiratory function must include both the pumping apparatus and the lung, respiratory function safety studies are best designed to evaluate both of these functional components. The total respiratory system is evaluated first by testing for drug-induced changes in ventilatory patterns of intact conscious animals. This is followed by an evaluation of drug-induced effects on the mechanical properties of the lung in anesthetized or paralyzed animals. Together, these evaluations are used to determine (1) if drug-induced changes in the total respiratory system have occurred, and (2) whether these changes are related to pulmonary or extrapulmonary factors.

6.2.2   Study design

The time intervals selected for measuring ventilatory patterns following oral administration of a drug should be based on pharmacokinetic data. The times selected generally include the time to reach peak plasma concentration of drug (T_{max}), at least one time before and one after T_{max}, and one time that is approximately 24 h after dosing to evaluate possible delayed effects. If the drug is given as a bolus intravenous injection, ventilatory parameters are monitored for approximately 5 min pre-dose and continuously for 20–30 min post-dose. Time intervals of 1, 2, 4, and 24 h are also monitored to evaluate possible delayed effects. If administered by inhalation or intravenous infusion, ventilatory parameter would generally be monitored continuously during the exposure period and at 1, 2, 3, and 24-h time intervals after dosing.

The time interval showing the greatest ventilatory change is selected for evaluating lung mechanics. However, if no ventilatory change occurred, the T_{max} would be used. If the mechanical properties of the lung need to be evaluated within 30 min after dosing, then dynamic measurements of compliance and resistance are performed. Measurements include a pre-dose baseline and continuous measurements for up to approximately 1 h post-dose. If the mechanical properties of the lung need to be measured at 30 min or longer after dosing, then a single time point is selected and the pressure-volume and flow-volume maneuvers are performed.

Supplemental studies including blood gas analysis, end-tidal CO_2 measurements, or responses to CO_2 gas and NaCN can be conducted to gain mechanistic insight after the ventilatory and lung mechanical findings have been evaluated. In general, these would be conducted as separate studies.

6.2.3 Capnography

The measurements of rates, volumes, and capacities provided by plethysmograph measurements have a limited ability to detect and evaluate some ventilatory disorders [16,25].

Detection of hypo- or hyperventilation syndromes requires measurement of the partial pressure of arterial CO_2 ($PaCO_2$). In humans and large animal models this can be accomplished by collecting arterial blood with a catheter or needle and analyzing for $PaCO_2$ using a blood gas analyzer. In conscious rodents, however, obtaining arterial blood samples by needle puncture or catheterization during ventilatory measurements is generally not practical. An alternative and noninvasive method for monitoring $PaCO_2$ is the measurement of peak expired (end-tidal) CO_2 concentrations. This technique has been successfully used in humans [33] and recently has been adapted for use in conscious rats [34]. Measuring end-tidal CO_2 in rats requires the use of a nasal mask and a microcapnometer (Columbus Instruments, Columbus, OH) for sampling air from the mask and calculating end-tidal CO_2 concentrations. End-tidal CO_2 values in rats are responsive to ventilatory changes and accurately reflect changes in $PaCO_2$ [34].

Our laboratory has developed a noninvasive procedure in conscious rats that is used to help distinguish between the central and peripheral nervous system effects of drugs on ventilation. Exposure to CO_2 gas stimulates ventilation primarily through a central mechanism [2,3]. In contrast, a bolus injection of NaCN produces a transient stimulation of ventilation through a mechanism that involves selective stimulation of peripheral chemoreceptors [24]. Thus, to distinguish central from peripheral nervous system effects, our procedure measures the change in ventilatory response (pretreatment versus posttreatment) to both a 5-min exposure to 8% CO_2 gas and a bolus intravenous injection of 300 ug/kg of NaCN. A central depressant (e.g., morphine sulfate) inhibits the CO_2 response and has little effect on the NaCN response.

6.3   Study design considerations

6.3.1   Dose selection

Safety pharmacology studies are designed to evaluate acute functional changes and thus involve single administration of test drugs. Further, since pharmacologic and toxicologic effects can be route specific, the test drugs are given by the intended clinical route. To evaluate dose response relationships, two or more doses are used. Selection of the high dose is based on acute toxicologic findings and is generally the lowest dose that produces evidence of toxicity (minimally toxic dose). This dose, however, must not be associated with toxicologic changes in secondary organs systems that may compromise the respiratory function measurements. The middle and low doses are generally log or half-log decrements of the high dose, with the low dose no less than the primary pharmacologic or anticipated human dose.

6.3.2   Species selection

The species selected for use in safety pharmacology studies should be the same as those used in toxicology studies. The advantages of using these species (rat, dog, or monkey) are (1) the pharmacokinetic data generated in these species can be used to define the test measurement intervals, and (2) acute toxicity data can be used to select the appropriate high dose. Further, the toxicologic/pathologic findings in these species can be used to help define the mechanism of functional change. The rat is the primary choice since rats are readily available, and techniques for measuring pulmonary function are well established in this species.

6.4   Summary

The known effects of drugs from a variety of pharmacologic/therapeutic classes on the respiratory system and worldwide regulatory requirements support the need for conducting respiratory function evaluations in safety pharmacology. Safety pharmacology studies of the respiratory system should include evaluations of both the total respiratory system and the mechanical properties of the lung. Changes in ventilatory patterns of conscious animals are used to evaluate the total respiratory system and the parameters measured include respiratory rate, tidal volume, minute volume, peak (or mean) inspiratory flow, peak (or mean) expiratory flow, and fractional inspiratory time. Changes in lung airflow and elasticity are used to evaluate the mechanical properties of the lung. The functional endpoints used to evaluate lung airflow include PEF (peak expiratory flow), FEF_{25} (forced expiratory flow-25%), FEF_{75}, FEV (forced expiratory volume), and dynamic resistance (or conductance); while the endpoints used to evaluate lung elasticity include TLC (total lung capacity), IC (inspiration capacity), FRC (forced respiratory capacity), and compliance. The species recommended for use in safety pharmacology studies are those used in toxicology studies since pharmacokinetic and toxicologic/pathologic data would be available. These data can be used to help select test measurement intervals and doses and to aid in the interpretation of functional change. The techniques and procedures for measuring respiratory function parameters are well established in guinea pigs, rats, and dogs.

References

1. McNeill, R.S., Effect of a β-adrenergic blocking agent, propranolol, on asthmatics, *Lancet*, 1964, Nov., 21:1101–1102.
2. Touvay, C., and Le Mosquet, B., Systeme respiratore et pharmacology de securite, *Therapie*, 2000, 55:71–83.
3. Borison, H.L., Central nervous system depressants: control-systems approach to respiratory depression, *Pharmacol. Ther. B.*, 1977, 3:211–226.
4. Akoun, G.M., et al., Natural history of drug-induced pneumonitis. In: *Drug Induced Disorders, Volume 3: Treatment Induced Respiratory Disorders* (Akoun, G.M., and White, J.P., eds.), New York: Elsevier Scientific Publishers B.V., 1989, pp. 3–9.
5. Dorato, M.A., Toxicological evaluation of intranasal peptide and protein drugs, *Drugs and the Pharma. Sci.*, 1994, 62:345–381.
6. Lalej-Bennis, D., et al., Six month administration of gelified intranasal insulin in 16 type 1 diabetic patients under multiple injections: efficacy vs. subcutaneous injections and local tolerance, *Diabetes and Metab.*, 2001, 27:372–377.
7. Gardner, D.E., *Toxicology of the Lung*, 4th ed. Boca Raton, FL: Taylor & Francis, 2006.
8. Rosenow, E.C., et al., Drug-induced pulmonary disease. An update, 1992, *Chest*, 102:239–250.
9. Eldridge, F.L., and Millhorn, D.E., Central regulation of respiration by endogenous neurotransmitters and neuromodulators, *Ann. Rev. Physiol.*,1981, 3:121–135.
10. Keats, A.S., The effects of drugs on respiration in man, *Ann. Rev. Pharmacol. Toxicol.*, 1985, 25:41–65.

11. Mueller, R.A., et al., The neuropharmacology of respiratory control, *Pharmacol. Rev.*, 1982, 34:255–285.

12. Tattersfield, A.E., Beta adrenoreceptor antagonists and respiratory disease, *J. Cardio. Pharma.*, 1986, 8(Suppl 4):535–539.

13. Shao, Z., Krishnamoorthy, R., and Mitra, A.K., Cyclodextrins as nasal absorption promoters of insulin-mechanistic evaluations, *Pharmaceutical Res.*, 1992, 9:1157–1163.

14. Shao, Z., and Mitra, A.K., Nasal membrane and intracellular protein and enzyme release by bile salts and bile salt fatty-acid mixed micelles-correlation with facilitated drug transport, *Pharmaceutical Res.*, 1992, 9:1184–1189.

15. Cherniack, N.S., Disorders in the control of breathing: hyperventilation syndromes. In: *Textbook of Respiratory Medicine* (Murray, J.F., and Nadal, J.A., eds.), Philadelphia: W.B. Saunders Co., 1988, pp. 1861–1866.

16. Gad, S.C., Safety assessment of therapeutic agents administered by the respiratory tract. In: *Toxicology of the Lung*, 4th ed. (Gardner, D.E., ed.), Boca Raton, FL: Taylor & Francis, 2006, pp. 231–296.

17. Milic-Emili, J., Recent advances in clinical assessment of control of breathing, *Lung*, 1982, 160:1–17.

18. Boggs, D.F., Comparative control of respiration. In: *Comparative Biology of the Normal Lung*, vol I. (Parent, R.A., ed.), Boca Raton, FL, CRC Press, 1992, pp 309–350.

19. Sinnet, E.E., et al., Fast integrated flow plethysmograph for small mammals, *J. Appl. Physiol.*, 1981, 50:1104–1110.

20. Leigh, D.E., and Mead, J., *Principles of body plethysmography*, National Heart and Lung Institute, Bethesda, MD: NIH, 1974.

21. Mauderly, J.L., The influence of sex and age on the pulmonary function of the beagle dog, *J Gerontol.*, 1974, 29:282–289.

22. O'Neil, J.J., and Raub, J.A., Pulmonary function testing in small laboratory mammals, *Environ. Health Perspect.*, 1984, 53:11–22.

23. Heymans, C., Action of drugs on carotid body and sinus, *Pharmacol. Rev.*, 1955, 7:119–142.

24. Heymans, C., and Niel, E., The effects of drugs on chemoreceptors. In: *Reflexogenic Areas of the Cardiovascular Systems* (Heymans, C., Neil, E. eds.), London: Churchill, Ltd., 1958, pp. 192–199.

25. Murphy, D.J., Safety pharmacology of the respiratory system: techniques and study design, *Drug Dev. Res.*, 1994, 32:237–246.

26. Amdur, M.O., and Mead, J., Mechanic of respiration in unanesthetized guinea pigs, *Am. J. Physiol.*, 1958, 192:364–368.

27. Diamond, L., and O'Donnell, M., Pulmonary mechanics in normal rats, *J. Appl. Physiol.: Respir. Environ. Exercise Physiol.*, 1977, 43:942–948.

28. King, T.K.C., Measurement of functional residual capacity in the rat, *J. Appl. Physiol.*, 1966, 21:233–236.

29. Illum, L., Davis, S.S. (reprint), Intranasal insulin: clinical pharmacokinetics, *Clinical Pharmacokin.*, 1992, 23:30–41.

30. Brown, L.K., and Miller, A., Full lung volumes: functional residual capacity, residual volume, and total lung capacity. In: *Pulmonary Function Tests: A Guide for the Student and House Officer* (Miller, A. ed.), New York: Grune & Stratton, Inc., 1987, pp. 53–58.

31. American Thoracic Society, Lung function testing: selection of reference values and interpretative strategies, *Am. Rev. Respir. Dis.*, 1991, 144:1202–1218.

32. Drazen, J.M., Physiological basis and interpretation of indices of pulmonary mechanics, *Environ Health Perspect.*, 1984, 56:3–9.
33. Nuzzo, P.F., and Anton, W.R., Practical applications of capnography, *Resp. Ther.*, 1986, 16:12–17.
34. Murphy, D.J., Grando, J.C., and Joran, M.E., Microcapnometry: a non-invasive method for monitoring arterial CO_2 tension during ventilatory measurements in conscious rats, *Toxicol. Mech. Methods*, 1994, 4:177–187.
35. Palecek, F., Measurement of ventilatory mechanics in the rat, *J. Appl. Physiol.*, 1969, 27:149–156.

chapter seven

Renal function

While the kidney is not included in the *pro forma* list of three mandatory organ systems that must be assessed before entry of a new pharmaceutical into humans, this is merely because interruption of its function is not immediately lethal [1]. The kidney is a very important and the second most common target for toxicity of numerous classes of drugs (Tables 7.1 and 7.2); however, unfortunately, modest adverse renal pharmacologic effects may elude detection by classical regulatory toxicological studies—but those same effects may be critical in the many patient populations with impaired renal function [4]. It is thus essential for safety pharmacologists to evaluate the potential for such effects before entry into man if a drug is intended for use in an implant; such detections require use of the full range of validated functional evaluation methods that are well established (Table 7.3).

It is mandatory to select accurate, clinically relevant parameters to be in a position to detect putative nephrotoxic effects during the safety pharmacology program. The glomerular filtration rate appears to be of major interest since it is associated with the definition of acute renal failure. Measurement of the renal blood flow, proteinuria, enzymuria, fractional excretion of sodium, the primary nephrotoxicity markers [5], and so forth are also highly useful to detect any possible renal impact of a new compound. Although the rat is by far the most widely used animal species, there are no specific (clinically relevant) reasons to choose it. Various parameters may vary according to the species, sex, strain, age, and other factors. Since in most cases acute renal failure occurs following administration of drugs in patients with pre-existing risk factors, it is suggested that sensitized animal models be validated and used (e.g., salt depletion, dehydration, co-administration of pharmacologic agents).

The kidney is a privileged target for toxic agents because of the following:

Physiological Properties:
- It is the most irrigated organ per gram of tissue (~400 ml/100g) and therefore more exposed to exogenous circulating toxins than many other organs.
- Tubular mechanisms of ion transport act to facilitate drug entry into renal tubular cells.

Table 7.1 Incidence of Drug-Induced Acute Renal Failure

Renal failure (high incidence >3%)

Amphotericin B, arsenic trioxide, mitoxantrone, levocarnitine, mycophenolate, mofetil, tacrolimus, naproxen, contract agents

Renal failure (low incidence <3%)

>150 different pharmaceuticals, including sumatripan, aspirin, indomethacin, aprotinin, tretinoin, propofol, graftskin, bivalirudin, rasburicase, hemophilus B conjugate, allopurinol, albenzadole, biphosphonates

Renal failure, neonatal (low incidence <3%)

Amlodipine, benazepril, captopril, enalapril, felodipine, fosinopril, lisinopril, moexipril, perindopril, quinapril, ramipril, losartan

Table 7.2 Example of Nephrotoxic Drugs [2,3]

- Anitbiotics: aminoglycosides, sulfonamides, methicillin, cephaloridin, polymyxins, etc.
- Nonsteroirdal anti-inflammatory drugs
- Iodinated contrast media
- Immunosuppressive drugs
- Angiotensin-converting enzyme inhibitors
- Chemotherapeutic agents (cisPT, methotrexate, etc.)
- Heavy metals (inorganic Hg + salts, Cd, Fe, As, Bi, Th...)
- Fluorinated anesthetics
- Dextrans

Pharmacokinetic Properties:
- Involved in filtration, secretion, and reabsorption of drugs.
- Kidneys concentrate urine and therefore intratubular drug concentration may be much higher than plasma concentration.
- High metabolic rate.

7.1   *Major functions of the kidney*

1. To maintain a stable chemical and physical environment for cells (this entails regulation of water, electrolytes, and acid/base rations)
2. To excrete metabolic wastes, including
   - uric acid, derived from nucleic acids
   - creatinine from muscle creatine
   - bilirubin, derived from hemoglobin
   - urea nitrogen derived from dietary and endogenous proteins
3. To excrete many foreign chemicals, including drugs
4. To regulate arterial blood pressure (Na^+ and water balance, rennin/angiotensin/aldosterone system, vasoactive PGs, etc.)

Table 7.3 Renal Function Endpoints

Excretory functions
Urine volume
S.G./osmolality
Na/K ratio
Creatinine (Cr)
Urea
Glucose, amino acids
Proteins
Crystals
Blood (hemolysis)
Biomarkers
Renal dynamics
Glomerular filtration rate (GFR)
Renal blood/plasma flow (RBF, RPF)
Urine flow rate
Electrolyte/solute excretion (H, Li, Na, K, CL, Ca, MG, PO_4, NH_4, urea)
Fractions filtered/reabsorbed
Osmolar clearance, CH_2O
Segmental analyses

5. To produce the active form of vitamin D
6. To degrade or excrete hormones (gastrin, insulin, PTH, GH, CCK, ADH, secretin, glucagons, etc.)
7. To be capable of gluconeogenesis during a prolonged fast
8. To release renal erythropoeitic factor, which acts on a liver globulin to produce erythropoietin, in case of decreased O_2 delivery to the kidney

Glomerular filtration rate (GFR) is the best global estimate of renal function. Insulin clearance is the gold standard for GFR measurement. Creatinine clearance may also be used, although its value exceeds the exact value for GFR because creatinine is not exclusively excreted by the tubules but also secreted by the tubules. This discrepancy increases as GFR falls. Serum creatinine measurement does not reflect abnormal renal function until after GFR has been reduced to at least 50% of the baseline value [6].

Other parameters are of great interest for a global approach of renal safety: fractional excretion of Na^+ and K^+, renal blood flow (*p*-aminohippuric acid clearance or ultrasonic transit-time flowmetry), enzymuria (which could allow differential location of toxic injuries), proteinuria,

Table 7.4 Risk Factors for Drug-Induced ARF [8,9]

Patient-related risk factors

Age, sex, race

Preexisting renal insufficiency

Specific diseases (diabetes mellitus, multiple myeloma, lupus, diseases
  associated with proteinuria, etc.)

Sodium-retaining states (cirrhosis, heart failure, nephrosis)

Dehydration and volume depletion

Hyperuricemia, hyperuricosirua

Sepsis, shock

Renal transplantation

Drug-related risk factors

Inherent nephrotoxic potential

Dose

Duration, frequency, and form of administration

Repeated exposure

Drug interaction

Associated use of diagnostic or therapeutic drugs with added or synergistic
  nephrotoxic potential.

glucosuria, diuresis, concentrating ability of the kidney (measurement of urine osmolality), and others.

Under normal conditions, GFR is submaximal because adaptive increases in single nephron GFR follow loss of damaged nephrons. Sensitized animal models that mimic risk factors commonly found in patients with drug-induced acute renal failure are advisable. This need should stimulate research in the field of safety pharmacology. The choice of the species, strain, and sex of test animals should take into account physiological and/or pharmacotoxicological specificities.

7.2   Acute renal failure

Acute renal failure (ARF) is the deterioration of renal failure over a period of hours to days, resulting in the failure to excrete nitrogenous waste products and to maintain fluid and electrolyte homeostasis.

ARF due to toxic or ischemic injury is the clinical syndrome referred to as acute tubular necrosis: common disease with high overall mortality (approximately 50%). Little progress has been made in treatment since the advent of dialysis more than 30 years ago, [7]. Table 7.4 summarizes the known risk factors for ARF, which should always be kept in mind during drug development (especially in clinical trials).

Table 7.5 ICH Renal System Primary Endpoints

Renal Function—Measurement of effects on urine excretion in saline-loaded rats
Renal Dynamics—Measurement of renal blood flow, GFR, and clearance

7.3 Functional reserve of the kidney

Adaptive increases in single nephron GFR tend to obscure renal injury until a considerable amount of kidney parenchyma is irreversibly lost. Under normal conditions, GFR is submaximal. Because of the marked amount of functional reserve in the kidney, no decrease is seen in impaired mammal animals until significant impairment has occurred. Such impairments, while not detectable in normal populations, can be quite significant in those that already suffer from renal disease; and as a result, consideration should be given to use animals that have been made more sensitive to functional effects by having their existing capacity reduced. Means of doing this would include surgical removal of one kidney as a measured renal toxic insult [10,11].

Table 7.5 presents the basic guidance provided by the International Conference on Harmonisation of Technical Requirements for Registration of Pharmaceuticals for Human Use (ICH) for renal safety pharmacology evaluation of new drugs. These evaluations are noncore and therefore not required prior to first-in-man studies. It is essential that intact animals be used in such evaluations, but it must be kept in mind that there are species differences in responses [12,13] and that renal function is readily influenced by anesthesia [14,15].

7.4 Clearance

Clearance techniques provide a convenient quantitative measure of renal function and are usually based on simultaneous analysis of plasma and urine. The clearance concept possesses the great advantage of permitting assessment of renal function in terms of a physiologically meaningful quantity, the virtual volume of plasma cleared of a solute by the kidney in unit time. The significance of the term *virtual* here lies in the fact that the same clearance value would be calculated whether a solute is completely extracted from 100 ml plasma per unit time, or only 50% cleared from 200 ml. Clearance, in other words, only corresponds to a discrete physical volume if a single pass through the kidneys leads to complete removal of the solute under study.

Conventionally, renal clearance has been equated to urinary clearance. In this case, the amount of the plasma solute excreted in urine per minute is given by the product of urine flow V (in ml/min) and the concentration of the solute in urine (U). If P denotes the steady concentration of that solute in arterial plasma, then the expression UV/P describes the volume of plasma from which the solute has been extracted into the urine over a period of 1 min. This defines the urinary clearance (C) of a solute as summarized by

$$C(ml/min) = UV/P$$

Strictly speaking, of course, renal accumulation or metabolism of a plasma solute also constitutes renal clearance from plasma. Thus, the kidney readily filters plasma Cd-metallothionein but almost completely reabsorbs it at low concentrations; little is excreted in urine [16], and the reabsorbed protein is not returned to blood. A more general expression for renal clearance from plasma clearance in ml/min is therefore given by the product of renal plasma flow (RPF) and E, the percent extraction calculated as (A–V)/A, where A and V represent the arterial and venous plasma concentrations of the solute under study. Note that the simplifying assumption is implicitly made that solute fluxes between red cells and plasma do not affect A or V. The general formula for clearance is thus

$$C = E \times RPF$$

Measuring clearance by determination of renal plasma flow and the A–V difference is, of course, more invasive and less convenient than basing the calculation, when possible, or urinary excretion. In further discussion of clearance, we shall therefore restrict ourselves to renal clearance as classically defined in terms of urinary excretion of plasma solutes. Note, therefore, that the clearance concept cannot be applied to urinary excretion of a solute like ammonia, which is synthesized in the kidney, or cadmium, whose excretion may result in part from release of metal sequestered for long periods of time in the kidney. Always keep in mind that only if the solute were completely removed from plasma during a single passage through the kidneys, that is, only if E as described above equaled 100%, with subsequent excretion into the urine, would the calculated clearance correspond to a specific physiological parameter, namely the renal plasma flow. However, even for solutes like *p*-aminohippuric acid (PAH) that most closely approach these conditions, extraction from plasma seldom exceeds 90%.

The clearance of an inert solute like insulin, which is freely filterable at the glomerulus, and which is neither secreted nor reabsorbed in the tubules, by definition must equal the GFR. Note the demand for complete filterability of the quantity of solute expressed by P in the initial equation. Thus, a major fraction of heavy metals in plasma is bound to plasma protein and can therefore not be filtered; accordingly, P in the expression UV/P must be corrected for nonfilterable metal for the clearance calculation to lead to a physiologically meaningful result. In practice, ultrafilterability of a solute circulating in plasma may be difficult to measure: First, properties of pores in synthetic ultrafiltration membranes may differ greatly from those of glomerular pores; a different charge distribution *in vitro* is likely to alter the filtration characteristics of ionic macromolecules from those observed *in vivo*. A second difficulty, sometimes overlooked, is related to the fact that anticoagulants, for instance by chelating heavy metals, may alter their filterability [17,18].

Renal clearance of extracellular solutes that are not otherwise excreted or metabolized can be calculated from their plasma disappearance curves [19]. Such techniques routinely utilize isotopically labeled compounds and possess the advantage of not requiring either urine collections or prolongd equilibrating infusions. When urinary clearance is conventionally determined from the U/P ratio, the arterial plasma concentration of the solute should ideally have reached a steady state. For an exogenous solute, the classical clearance technique therefore requires as long as 1 h of equilibrating infusion, often preceded by a priming injection. Alternatively, and more conveniently, reasonably stable plasma levels of solutes can be achieved by giving an intramuscular or subcutaneous depot injection. Even if absolute constancy is not achieved in this manner, a mean concentration is slow and constant.

The steady-state solute concentration, however achieved, must always lie well below the half-saturating concentration (K_m) of any nonlinear reabsorptive or secretory process in the renal tubule. A well-known illustration of this fact is the glucosuria of diabetes mellitus, which reflects not damage to the proximal tubules but saturation of their capacity to transport the excess sugar. Another prerequisite for significant clearance values is a urine flow sufficiently high to permit essentially complete collections during accurately timed clearance periods. Even carefully controlled experiments in the laboratory cannot always achieve such ideal conditions. As a result, renal clearance values seldom possess confidence limits of better than perhaps +10%. Such accuracy may be adequate for most purposes; it implies, however, that as much as 20% depression of a renal solute clearance following exposure to a compound believed to act directly on the kidney does not necessarily indicate a nephrotoxic effect.

A special case is that of endogenous creatinine. This is normally produced at a constant rate as the end product of muscle metabolism and therefore circulates in plasma at a constant concentration; it is freely filtered at the glomerulus but neither reabsorbed nor effectively secreted in the tubule. Secretion has been reported in some species, but it is sufficiently small so that it does not significantly contribute to creatinine excretion except at low filtration rates. Like insulin clearance, that of creatinine can therefore serve as a convenient measure of the glomerular filtration rate. Creatinine possesses a great advantage over insulin in that its clearance can be measured without the need of an equilibrating infusion [18,19].

Measurement of the renal clearance of an exogenous solute by conventional techniques consumes relatively large amounts of this solute to reach and maintain the required steady plasma levels; this is especially true of determinations of maximum tubular transport activity T_m at high plasma concentrations (see the section "Measurement of T_m," below). Factors of cost of the solute, time required for its equilibration, toxicity, radiation exposure in the case of labeled solutes, and metabolic breakdown or transformation, as for instance by transamination of an infused amino acid into a competing amino acid, all constitute disadvantages or normal T_m measurements.

These problems can be minimized by use of a so-called arterial gradient infusion procedure in which the arterial plasma concentration of the solute in strongly diuresing animals is rapidly raised in a stepwise fashion. It can then usually be kept approximately constant for several minutes by decreasing the rate of infusion in a manner equal to and opposite of the rate at which recirculation would have raised plasma levels at a constant infusion rate. It is possible in this manner to carry out a complete clearance determination in 3 to 4 min [20]. The procedure has been successfully applied to dogs, rabbits, and rats but has not been widely adopted, perhaps because of occasional difficulties encountered in maintaining constant plasma concentrations.

Clearance determinations discussed so far all require measurement of concentrations in carefully timed urine and plasma samples. In addition, useful approximations to relative solute clearance values can be obtained by simplified procedures. The best known of these simply takes the plasma level of urea, or preferably creatinine, as a measure of the GFR. Indeed, if creatinine excretion (UV in g/day) is constant, the GFR (=UV/P) theoretically is inversely proportional to P_{Cr}, the creatinine concentration is plasma; any increase in P_{Cr} above a normal level of around 1 mg/dl should therefore reflect a corresponding fall in GFR. In practice the method is not very sensitive in the normal range of plasma creatinine levels (<1.4 mg/dl); a better correlation between measured creatinine clearance (C_{Cr}) and that predicted on the basis of P_{Cr} is obtained at higher plasma levels, that is, lower C_{Cr} values [6,21].

Because its rate of production is influenced by changes in protein metabolism and its excretion by the rate of urine flow, the level of urea in plasma varies more widely than that of creatinine. Although it is therefore a less reliable indicator of glomerular function than is plasma creatinine, plasma urea is frequently used to follow changes in GFR. Note that blood and plasma or serum urea are used interchangeably because this solute freely diffuses into and out of blood cells.

Considerable information can also be obtained on changes in clearance values by analysis of only randomly collected urine. This approach is also based on the characteristics of creatinine excretion as discussed above and consists of normalizing excretion of a solute by that of creatinine. A change in the solute/creatinine concentration ratio in urine then reflects a corresponding change in the excretion of the solute under study.

Normalization of urinary solute excretion on the basis of creatinine (or, less frequently, on the basis of urinary specific gravity) possesses other advantages. Thus, normalization significantly diminishes the large fluctuations often encountered in urinary solute concentrations. It also justifies, as discussed above, the analysis of spot samples of urine, thus avoiding the inconvenience of having to collect 24-hr urine samples. Because of diurnal variations in the GFR, the spot samples should be collected at the same time each day.

Gross changes in renal plasma flow, or a reduction in the efficiency (E) of extraction of secreted solutes from blood due to cytotoxic effects, can also be demonstrated by changes in the time required for excretion of a standard injected dose of compounds like ortho-iodohippurate, the dye phenol red (PSP), and x-ray contrast media like Diodrast. Each of these substances is efficiently extracted by the normal kidney, with a high value of E, so that their excretion provides a composite measure of effective renal plasma flow, total tubular mass, and the functional integrity of the tubule cells. There are now some elegant methods for measuring renal blood flow, such as ultrasound flowmetry [22]. Additional information on changes in tubular mass and activity can be obtained from determination of maximum tubular capacity T_m as discussed below.

A rapid and powerful approach to the study of renal function in control or poisoned animals is the double indicator dilution technique, introduced into renal studies by Chinard [23]. It consists essentially of a rapid arterial bolus injection of a test solute, together with a glomerular and extracellular marker such as creatinine, followed by rapid collection of sequential urine or renal venous blood samples. Although the procedure is too invasive for routine or other-than-terminal function assessment, it can provide extensive information on the integrity of solute transport processes and on their mechanisms, as well as on the site of action of various inhibitors.

The indicator dilution technique possesses the disadvantage of requiring arterial injection and ureteral cannulation in the anesthetized animal. On the other hand, it is rapid and consumes relatively little solute, thus permitting convenient and repeated measurements of relative clearances. Care must be taken in all renal transit experiments to use sufficiently low bolus concentrations of the test solute so that even at its peak plasma concentration renal transport mechanisms do not become saturated.

Saturation of a reabsorptive system will, by definition, increase the fractional excretion of its substrate. As an example of the important use to which such observations can be put, reference may be made to the report that at high bolus concentrations of metallothionein (MT) the characteristics of its transit from artery to urine approach those of insulin. The implications of this finding are that MT is freely filterable and that it is normally reabsorbed by a saturable mechanism. In addition, obviously little nonfiltered MT can have been transported directly from blood into urine (secretion), a conclusion further confirmed by the similar renal vascular transit curves of insulin and of MT.

7.5   Free water clearance and renal concentrating ability

One of the most important functions of the normal kidney is its ability to respond to changes in the state of hydration by excreting dilute or concentrated urine. Appropriate water excretion or retention provides a very sensitive measure of general renal integrity.

The simplest measure of the ability of the kidney to conserve water is provided by osmolality of urine excreted by a water-deprived subject. The osmolality of plasma changes relatively little and averages close to 300 mOsm; analysis of the osmolar concentration of urine thus leads directly to the (U/P) osmol ratio. Simple and convenient osometers are available for such determinations. Human urine may range from osmolalities below 100 mOsm to a maximum of perhaps 1800, that is, to a urine sixfold more concentrated than plasma. The rat, in contrast, can readily concentrate its urine nine- to tenfold.

A more quantitative measure of urine concentration ability is the clearance of free water (C_{H_2O}), derived in turn from the osmolal clearance, (UV/P) osmol. The latter in effect represents the volume that would be required to excrete total urinary solute in isosmotic solution, that is, at a concentration of 300 mOsm. If filtrate were excreted as such, the osmolal clearance would equal the GFR. If the urine is more dilute than the filtrate, the dilution may be visualized as that volume of filtrated from which solute was removed without reabsorption of water; this volume is defined as the free water clearance (C_{H_2O}), or free water excretion, and is given below, where V stands as usual for the urine volume excreted per unit time:

$$C_{H_2O} = V - (UV/P)osmol$$

Dilution of urine, that is, reabsorption of solute without water, primarily occurs in the so-called diluting segment in the thick ascending loop of Henle. The water permeability of this segment is very low, and a NaCl is reabsorbed when the remaining fluid is rendered dilute. Additional solute reabsorption occurs in the more distal nephron and contributes to a high interstitial solute concentration in the medulla. In the presence of antidiuretic hormone, water can move out of the lumen along the osmotic gradient, so that the final osmality of the urine reflects the osmolal concentration in the medulla. Excretion of concentrated urine, by analogy with the formation of dilute urine discussed above, implies that a certain volume of solute-free water must have been removed from the isosmotic filtrate, in addition to the large fraction of filtered water and solute normally reabsorbed iso-osmotically in the proximal tubule. In this case, the clearance of solute-free water is negative, and approximately referred to as negative free water clearance; it is calculated as (UV/P) osmol-V.

Several intrarenal factors are involved in the production of concentrated urine, including especially the attainment and maintenance of high interstitial osmolality in the medulla as a result of solute filtration and reabsorption, and of the countercurrent concentrating mechanism, which depends in turn on normal medullary blood flow. Urinary concentrating ability can therefore not be associated with a specific aspect of renal function but serves rather as reflection of its general integrity, and can thus prove useful for screening purposes.

Two standard materials are used in the evaluation of clearance: insulin (which is freely filtered and neither reabsorbed nor secreted into the tubules) and creatinine (which is not exclusively filtered but also secreted into the tubules). Creatinine clearance may exceed the exact value for GFR, and this discrepancy increases as GFR falls. We may therefore consider the advantages and disadvantages of current clearance measurement methods in animals. (See Table 7.6 for a comparison of GFR in healthy and in diabetic rats, for example.) Current clearance measurement methods are performed in animals.

Advantages
- Can be used in about all species
- No need for anesthesia or surgery that may alter renal function
- Link between animal and human studies
- Functional approach

Disadvantages
- Cannot distinguish internephron variation
- Depends upon circadian changes in rats and humans

Table 7.6 GFR in Healthy and Diabetic Rats [24]

	Control rats		Diabetic rats	
	Conscious	Anesthetized	Conscious	Anesthetized
Insulin clearance ml/min	3.2 ± 0.2	1.8 ± 0.2	3.6 ± 0.2	3.1 ± 0.6
Insulin clearance ml/min/ kg BW	10.9 ± 0.05	6.8 ± 0.8	15.9 ± 0.8	10.9 ± 1.7

Note: Sprague Dawley female rats: BW 270–300 g; anesthesia: Inactin 100 mg/kg
Diabetic rats: streptozotocin, single dose 60 mg/kg, 2–4 months prior to the study

7.5.1   Renal blood flow

Plasma clearance of para-aminohippuric acid (PAH) is almost equal to a plasma flow.

PAH is secreted into the tubules on a single pass, and 10%–15% of the RPF goes to nonfiltering portions of the kidney. This nonfiltered plasma cannot therefore lose its PAH, and thus PAH clearance = effective renal plasma flow = 0.5%–90% of total RPF.

7.5.2   Fractional excretion of sodium

The percentage of filtered Na^+ load excreted in the final urine is a measure of renal function, as is also the clearance of Na^+ expresses as a percentage of GFR: FE_{Na} (%) = $(U_{Na} \times V) / (P_{Na} \times GFR) \times 100$.

Theoretically, if drug-induced damage of the medullopapillary portion of the nephron, decreases in reabsorption of Na^+ and water lead to increase in FE_{Na}.

7.6   Clinical chemistry measures

The most useful markers for changes in renal function and renal toxicity without special manipulation are clinical chemistry markers, including the newly identified set of seven identified by the Critical Path Initiative [5]. These can identify specific impaired function or toxicity (Table 7.7).

Table 7.7 Clinical Chemistry Measures of Specific Renal Effects

Origin or urinary enzymes	Enzymes
Brush border	alanine-aminopeptidase
	γ-glutamyl-transferase
	trehalase
Lysosomes	β-glucuronidase
	N-acetyl-β-D-glucosaminidase
	acid phosphatese
	β-galactosidase
Cytosol	lactate dehydrogenase
	leucine aminopeptidase
	β-glucosidase
	fructose-1,6 biphosphatase (proximal tubule)
	pyruvate kinase (distal tubule)

Enzymuria depends upon the following:

- Circadian and infradian rhythms
- Urinary flow rate
- Urinary pH
- Age
- Sex
- Environmental pollution

Examples of enzymuria include the following:

Proteinuria. In general, albuminuria leads to increases in the permeability of the glomerular capillary wall, whereas urinary release of low-molecular weight proteins (β-2-mg) suggest an impairment in tubular reabsorption.

Glucosuria. Normal serum glucose concentration may reveal proximal tubular damage (such as with gentamicin and maleic acid).

Urine Concentration Test. Food and water are withheld for 24 h. Osmolality of the collected urine is measured. Some toxic agents decrease maximal concentrating capacity (cis-dichlorodiammineplatinum, etc.). The Critical Path Initiative [5] has identified seven enzymatic markers (kidney injury molecule, albumin, total protein, β_2 microglobulin, Crystatin C, clustorin, and the trefoil factor) to reliably detect renal damage. These have now been accepted by all the ICH primary regulatory agencies.

7.7 Animal models

7.7.1 Rat

While there is no specific scientific reason to select the rat for evaluating renal safety, three factors should be kept in mind:

- Easy handling
- Availability of numerous strains with metabolic and/or physiological specificities: BBB; Munich-Wistar; Lewis-DA; hypertensive salt-sensitive Dahl strain, SHR, etc.
- Extensive historic database

When comparing GFR and RPF between rats of different strains, it is appropriate to correct these parameters for body weight because they are genetically correlated; but for high within-strain variation of kidney weight not correlated with renal function, don't correct GFR for kidney weight.

In Wistar rats [25], differences are as follows:

- Salt depletion: after four daily injections of furosemide (2 mg/kg), rats fed with boiled rice (Na⁺ 2 mg/100 g) for 7 d before test compound
- Indomethacin 10 mg/kg 1 h before test compound
- Uninephrectomy 3–5 weeks before
- Dehydration the day of test-compound injection

Another example: Dehydration greatly enhances the sensitivity of rats to the nephrotoxicity of aminoglycosides.

7.7.2 Dog

- Ethical limitations
- Easy surgical accessibility to kidney and vessels
- Single nephron GFR = 60 nl/min closer to that humans (60–65 nl/min) than that of rats (about 33 nl/min)
- Renal medulla = simple type with small, cone-shaped vascular bundles as in humans

Caution should be taken as follows:

- All anesthetics induce hemodynamic changes.
- Barbiturate can reduce GFR by 50% in rats.
- Thiobutabarbital (Inactin) impairs RBF autoregulation in dogs and rats, not pentobarbital.

7.8 Examples of species differences in drug sensitivity

Ethacrynic Acid
- Drastic diuretic in humans, monkey, dog, mouse
- Lesser effect in rats

Ouabain
- Effective inhibitor of Na⁺/K⁺ ATPase in humans, dog
- Less effective in guinea pig, rabbit, very low activity in rat

Amino-glycosides
- Higher nephrotoxic threshold in animals (10- to 60-fold) than in humans

Iodinated Contrast Media
- Less toxic in rats than man

References

1. ICH, *ICH S7A ICH Harmonized Tripartite Guidelines on Safety Pharmacology Studies for Human Pharmaceuticals*, 2001.
2. Fillastre, J.P., et al., Détection de la néphrotoxicité médiicamenteuse, *J. Pharmacol.*, Paris, 1986, 17(suppl. I):41–50.
3. Thatte, L., and Vaamonde, C.A., Drug-induced nephrotoxicity, *Postgrad. Med.*, 1996, 100:83–100.
4. CDER, FDA, *Guidance for Industry: Pharmacokinetics in Patients with Impaired Renal Function—Study Design, Data Analysis, and Impact on Dosing and Labeling, Draft Guidance*, Rockville, MD: FDA, March, 2010.
5. Hoffman, D., et al., Performance of novel kidney biomarkers in preclinical toxicity studies, *Toxicol. Sci.*, 2010, 116:8–22.
6. Cockroft, D.W., and Gault, M.H., Prediction of creatinine clearance from serum creatinine. *Nephron*, 1976, 16:31–41.
7. Brezis, M., Rosen, S., and Epstein, F.H., Acute renal failure. In: *The Kidney* (Brenner, B.M., and Rector, F.C., eds.), Philadelphia, PA.: WB Saunders, 1991, pp. 993–1061.
8. Thadani, R., Pascual, M., and Bonventre, J.V., Acute renal failure, *New Engl. J. Med.*, 1996, 334:1448–1460.
9. Lieberthal, W., and Nigam, S.K., Acute renal failure, II. Experimental models of acute renal failure: imperfect but indispensable, *Am. J. Physiol.*, 2000, 278:F1–F12.
10. Idée, J.M., Renal safety pharmacology: value of sensitised experimental models. *Thérapie.*, 2000, 55:91–96.
11. Lauwerys, R.R., and Bernard, A., Early detection of the nephrotoxic effects of industrial chemicals: state of the art and future prospects, *Am. J. Ind. Med.*, 1987, 11:275–285.
12. Chiu, P.J.S., Models used to assess renal function, *Drug Dev. Res.*, 1994, 32:247–255.
13. Craddock, G.N., Species differences in response to renal ischemia. *Arch. Surg.*, 1976, 111:582–584.
14. Conger, J.D., and Burke, T.J., Effect of anesthetic agents on autoregulation of renal hemodynamics in the rat and dog, *Am. J. Physiol.*, 1976, 230:652–657.
15. Buchardi, H., and Kaczmarczyk, G., The effect of anaesthesia on renal fucntion. *Eur. J. Anaesth.*, 1994, 11:163–168.
16. Nomiyama, I.K., and Foulkes, E.C., Reabsorption of filtered cadmium metallothionein in the rabbit kidney. *Proc. Soc. Exp. Biol. Med.*, 1977, 156:97–99.
17. Foulkes, E.C., Mechanisms of renal excretion of environmental agents. In: *Handbook of Physiology*, Section 9, "Reactions to Environmental Agents" (Lee, D.H.K., ed.), Washington, DC: American Physiological Society, 1977, 495–502.
18. Harvey, A.M., and Malvin, R.L., Comparison of creatinine and inulin clearances in male and female rats, *Am. J. Physiol.*, 1965, 209:849–852.
19. Tepe, P.G., et al., Comparison of measurements of glomerular filtration rate by single sample, plasma disappearance slope/intercept and other methods, *Eur. J. Nucl. Med.*, 1987, 13:28–31.
20. Foulkes, E.C., On the mechanism of chlorothiozide-induced kaliuresis in the rabbit. *J. Pharmacol. Exp. Ther.*, 1965, 406–413.

21. Jobin, J., and Bonjour, J.P., Measurement of glomerular filtration rate in conscious, unrestrained rats, *Am. J. Physiol.*, 1985, 248:F734–F738.
22. Evans, R.G., et al., Chronic renal blood flow measurement in dogs by transit-time ultrasound flowmetry, *J. Pharmacol. Toxicol. Meth.*, 1997, 38:33–39.
23. Chinard, F.P., Relative renal excretion patterns of p-aminohippurate (PAH) and glomerular substances, *Am. J. Physiol.*, 1956, 185:413–417.
24. Hierholzer, K., Value and necessity of animal research in nephrology, *Expl. Biol. Med.*, 1982, 7:88–101.
25. Heyman, S.N., et al., Acute renal failure with selective medullary injury in the rat. *J. Clinical Invest.*, 1988, 82:401–412.

chapter eight

The gastrointestinal system

Alterations in gastrointestinal (GI) function are common side effects of many drugs, not limited to those administered orally [1,2]. Changes in bowel habits (constipation or diarrhea) and gastric mucosal irritation are the most common side effects that are covered at least in part by classical technical approach involving the evaluation of gastric emptying, intestinal and colonic transit, as well as direct gastric mucosal damage score evaluation [3,4]. For example, increased small intestinal permeability caused by nonsteroidal anti-inflammatory drugs (NSAIDs) is probably a prerequisite for NSAID enteropathy, a source of morbidity in patients with rheumatoid arthritis. These results were supported by the 51Cr EDTA/L-rhamnose urine excretion ratios, which reflect changes in intestinal permeability [5]. Effects of drugs on intestinal tone has appeared recently to be of importance since compounds having a relaxatory effect on colonic muscular tone may favor the occurrence of ischemia. However, there is now evidence that other functions of the gut may be affected, giving rise to more chronic alteration and/or reactivity to oral pathogen or irritant or alteration in digestion and presence of gut hypersensitivity to mechanical stimuli. Several other gut functions such as epithelial barrier regulating both transcellular and paracellular absorption may be altered as well as enzymes, hormones, and ion secretion. Intestinal and colonic microflora may be also affected by orally administered drugs. Viscerosensitivity of the gut is affected by the immune status of the mucosa. There is accumulated evidence that orally administered drugs may modify this immune status giving rise to sensitization of sensory nerve terminals, but a direct action after epithelial absorption on receptors located on terminals of primary afferent neurons is possible. These effects may trigger abdominal pain.

Increased paracellular permeability such as that induced by the majority of nonselective NSAIDs favors the entry of pathogens, allergens, and bacterial translocation, giving rise to enteropathy and septic shock. Simple tests performed in animals by measurement of permeability to macromolecules in basal and simulated conditions may provide relevant information. Similarly, drugs may directly affect both functional and nociceptive sensitivity of the gut and subsequently inhibit enzyme secretion and many upstream inhibitory reflexes, and initiate abdominal pain in response to normal mechanical stimuli.

Not only antibiotics but other drugs orally administered may have an effect on the colonic microflora, here again altering the immune balance through the gut mucosa. Substances administered orally may also affect the population of immunocytes within the gut. Among them, mast cells are the most often affected population. This effect may be related or unrelated to sensitization through gut allergic reaction. Increased numbers of mast cells or of their contents may also contribute to sensitization of primary afferent terminals, here again affecting gut sensitivity. All this information suggests that it should be relevant to test the influence of new drugs on colonic sensitivity to distension through classical models used to evaluate the sensitivity of the gut to luminal distension.

Finally, safety pharmacology, particularly for drugs administered orally, also must evaluate the effect of drugs on other important functions of the gut with specific attention to paracellular permeability, enzyme secretion, immune status of the mucosa, and sensitivity.

8.1 Drug-induced alterations of GI transit or motility

- gastric emptying: →   dyspeptic symptoms
  - early satiety
  - postprandial fullness
  - regurgitations, nausea
- intestinal transit: →   irritable bowel syndrome (IBS) symptoms
  - abdominal fullness
  - gas retention
  - abdominal pain
- colonic transit: →   IBS symptoms
  - constipation
  - diarrhea
  - bloating, abdominal pain

Colonic tone: Association of reduced colonic muscular tone plus slowing of colonic transit may sometimes initiate ischemic colitis (e.g., 5-HT3antagonists-alosetron).

8.2 Gastrointestinal function

Following are GI functions that can be affected by drugs:

- Transit (motility)
- Absorption (transcellular and paracellular transport)
- Digestion (enzymes)
- Secretion (ions, hormones, enzymes)

Table 8.1 GI System

GI Function—Measurement of gastric emptying and intestinal transit
Acid Secretion—Measurement of gastric acid secretion (Shay rat)
GI Irritation—Assessment of potential irritancy to the gastric mucosa
Emesis—Nausea, vomiting

- Microflora (colon)
- Immunity (mucosa)
- Viscerosensitivity (abdominal pain)

The potential effects of new drugs on the digestive system can be examined in a number of model systems, of which intestinal motility in the mouse and/or gastric emptying in the rat are examples recommended for safety pharmacology evaluation. Intestinal motility, assessed by the transit of carmine dye in the mouse, and gastric motility, assessed by stomach weight in the rat, were examined using a range of clinical drugs or potent pharmacological agents known to affect gastrointestinal function. Assessment of both models in the guinea pig was also evaluated. Activity was demonstrated with codeine, diazepam, atropine, and CCK-8 (all of which inhibited gastric function). However, neither model gave consistent and reliable results with the remaining reference compounds, namely, metoclopramide, bethanechol, cisapride, deoxycholate, carbachol, and domperidone. In conclusion, this investigation questions the usefulness of simple models of gastrointestinal transport in the rodent as a means of detecting potential effects of a new drug on the digestive system. This finding should be of concern to the pharmaceutical industry as these simple models are routinely used as part of a regulatory safety pharmacology package of studies.

A number of classic assays have been designed to examine the effects of a test article on gastrointestinal function. Gastrointestinal transit rate is most often measured with a test employing a forced meal of an aqueous suspension of activated charcoal [6]. The test article is given via the appropriate route at a preset time prior to the charcoal meal, in the appropriate animal model [7]. For example, a compound intended for use via intravenous injection would be injected intravenously in mice 30 min prior to delivering a charcoal meal by gavage. The distance traveled from the stomach by the black-colored charcoal meal to a specific anatomic location within the intestine is measured at a fixed time after this meal, usually 20 or 30 min later. In validating this procedure at Mason Laboratories, the staff tested the ability of a parasympatholytic agent, intravenous atropine sulfate, to inhibit gastrointestinal transit [5]. In a dose-dependent fashion, 30 and 50 mg/kg atropine sulfate significantly decreased the distance traveled by the charcoal meal.

Appropriate methodology to evaluate effects on GI transit and motility include the following:

- Gastric emptying:
  - Nutritive radiolabeled meal
  - Solid and liquid phases
  - Rats, dogs
- Intestinal transit:
  - Radiolabeled meal
  - Determination of the geometric center for 10 equal intestinal segments
- Colonic transit:
  - Rats with intracolonic catheter
  - Accustomed to eat in a given time (3–4 h)
  - ^{51}Cr-Na administered intracolonically
  - Collection of feces by 30 min period during 24 h

8.3   Assessment of intestinal transit

A second major function of the intestinal tract depends on its contractility. Intestinal motility is responsible for appropriate mixing of ingested materials with endogenous secretions required for digestion and for appropriate delivery of substances to the site of absorption elimination.

One approach to the study of intestinal motility *in vivo* is the use of an intraluminal marker substance whose transit through the gut lumen can be quantitated [6]. Substances used as valid markers should not be absorbed from the intestine nor adsorbed onto the mucosal surface. In addition, ideal markers do not affect any aspect of intestinal function. Furthermore, since the contractile properties of the intestinal musculature may have different effects on solid as opposed to liquid components of the gut contents, markers should be chosen to correspond to the physical composition of the endogenous substance of greatest interest. This point is especially relevant to the study of gastric emptying [8]. Markers used in the tracing of solid substances include, for example, ^{99m}Techinitium incorporated into a chicken liver meal [8]. Markers such as this gamma emitter can be readily monitored for stomach content in humans using a gamma camera. Other solid markers, used primarily in analysis of total gastrointestinal transit time in humans, include radiopaque pellets made from polythene impregnated with barium sulfate [9].

A technique developed by Summer et al. [10] permits investigation of intestinal transit in animals by administering markers into the duodenum. This procedure allows interpretation of effects on intestinal transit without any influence of gastric emptying. Permanent, indwelling catheters are surgically implanted in the duodenum of rats and exteriorized behind the head. Markers such as ^{51}Cr can be administered directly into the small intestine without the disruptive effect of anesthesia or of oral intubation procedures. In animal studies the content of markers in the intestine can be determined by analyzing sequential gut segments after the animals have been killed.

Other approaches to the *in vivo* analysis of the motor function of the gut [11] include determination of intraluminal pressure changes. These changes can be measured using small balloons or fluid-filled open-tipped tubes connected to external strain gauges, or internal miniaturized strain gauges monitored by telemetry or by means of exteriorized wires [12]. Analysis of the function of the intestinal musculature may also entail measurement of its electrical activity, which in a chronic *in vivo* preparation can be carried out with a recording electrode surgically implanted on or in the intestinal wall or intubated into the intestinal lumen using a balloon to ensure its juxtaposition to the mucosal surface [13].

The contractile properties of intestinal smooth muscle can also be assessed using *in vitro* preparations of this organ. Such *in vitro* techniques are valuable in screening potentially toxic substances for effects on intestinal contractility and for elucidating mechanisms of effect on propulsion observed with *in vivo* methodology. Use of *in vitro* techniques to study intestinal motility has certain advantages over *in vivo* procedures. Generally, they are technically simpler to execute. They isolate the tissue from extrinsic neural and hormonal influences. The tissue can be directly exposed to the test substance. These advantages are at the expense of loss of prediction of *in vivo* effects of a test substance [14].

The choice of a particular *in vitro* technique depends on the specific aim of the experimentation. Those techniques that are most commonly used differ in several ways. First, the species from which the intestinal segment is taken markedly affects the basal contractile activity. The rabbit jejunum, for example, maintains rhythmic contractions *in vitro* and therefore is especially useful for analysis of substances suspected of having inhibitory effects on intestinal smooth muscle. The guinea pig ileum, in contrast, exhibits little spontaneous activity *in vitro*. This preparation is therefore widely used in the bioassay of agents causing contraction of intestinal smooth muscle. To test for depressant effects, the investigator must induce contraction of this tissue as with electrical stimulation. Second, there are differences in the responses of the smooth muscle, depending on the site within the intestine under investigation. This limits the investigator's ability to generalize from an experiment carried out with a muscle preparation from a single region of the intestine and reinforces the importance of strictly controlling the tissue region studied in a series of experiments.

8.4   Determination of intestinal absorption

Study of the absorptive function of the intestinal tract can be carried out with numerous methods. Among the primary considerations in choosing an experimental technique to assess an aspect of the absorption process are the following:

1. The test species to be used; that is, must the study be conducted in humans with all the accompanying complications, or is there an appropriate experimental animal model?
2. The aspect of the absorption process of interest; for example, is it the overall absorption from the gut lumen to the systemic circulation and tissues, or is it the process of transport across the brush border or basolateral membrane of the intestinal mucosal cell?
3. Which experimental or physiological variables should be controlled, for example, the presence of anesthetic agents, the electrochemical potential difference across the gut wall, or the pH of the luminal gut contents?

The answers to these questions will determine the particular method that may be chosen from among the *in vitro* methods available for studying absorptive function [15,16]. These methods can be categorized according to (a) the method by which the test substance is administered, and (b) the method for assessing the extent and/or rate of absorption.

8.4.1   Methods of administering test substance

Among the various techniques for administering the test drug in an *in vivo* study are the following:

1. Incorporating the test substance into the diet, which is then administered to the subject. This procedure may be especially relevant to the analysis of the absorption of drugs that are contaminants of the diet.
2. Intubating the test substance into the stomach, a procedure that allows more precise control of the total dose administered.

With both of the above methods the rate and possibly the extent of absorption of the test substance may be markedly affected by the gastric emptying pattern of the subject.

3. Directly administering the test substance into the intestinal lumen, which eliminates the influence of gastric emptying. In human studies, substances can be administered through small-bore intubating

tubes localized to particular sites by radiographic techniques [17]. In animal experimentation, test substances may be perfused through the gut lumen as a single pass, analogous to the perfusion method in man, or recirculated in the perfusate. Such a technique requires cannulation of the intestine, an external heating device for maintaining the perfusate at body temperature, and a pump for maintaining constant flow. An advantage of perfusion procedures, over techniques described below, is that the influence of flow rate on absorption kinetics can be directly determined. Analyses have indicated that for many substances, such as long chain fatty acids, bile acids, and cholesterol [18], diffusion through an unstirred water layer overlying the mucosal surface is a rate-limiting step in the overall absorption process.

Another method of direct administration in experimental animals consists of placing a test substance into a segment of the intestine that is closed by ligatures both proximally and distally [19]. The construction of the closed segment and the injection of the test dose do require the use of anesthetic agents. However, the animal, typically a small animal such as a rat, can be allowed to recover from anesthesia and to become ambulatory for the majority of the absorption period. This procedure has the additional advantage of not requiring perfusion pumps or heating devices.

Substances may also be administered directly into the intestinal lumen following surgical creation of exteriorized fistulas [15]. Studies are then carried out in unanesthetized larger animals, typically dogs. This approach, of historical importance, has the disadvantage of requiring considerable surgical manipulation.

8.4.2   Methods for quantitating degree of absorption

8.4.2.1   Appearance in systemic fluids

In addition to the method of administering the test substance, the second critical aspect of a technique is the sampling procedure for quantitating the extent and/or rate of absorption. With *in vivo* methods the least invasive techniques entail the collection of blood, urine, or breath samples for determining the appearance of the absorbed test substance (and its metabolites) in body fluids. Comparing the time course of plasma concentrations or excretory rates in urine or breath after oral administration with the results after intravenous administration may permit quantitation of the extent of absorption of the test substance and the rate constant of this process [20]. This approach is relatively imprecise and may be confounded by numerous factors such as the *first pass* effect, the enterohepatic circulation of the agent, and the status of elimination processes such as hepatic and renal function. Nevertheless, this technique is useful in the diagnosis of malabsorption syndromes associated with gastrointestinal diseases [21]. For example, one test of the transport capacity of the small intestine for carbohydrates entails oral administration of the pentose sugar D-xylose followed by its determination in plasma or in urine. Similarly, assessment of intestinal lactase, the disaccharide that cleaves lactose into the absorbable sugars, glucose and galactose, involves an oral lactose load followed by determination of blood sugars. A test of fat absorption can include determination of serum carotene. One test of ileal absorptive function entails the oral administration of radiolabeled vitamin B_{12} with determination of its urinary recovery.

Another approach for analyzing intestinal function is the sampling of excretory products in breath [22]. This technique entails analyzing breath for hydrogen, which is generated by the body exclusively by the action of intestinal bacteria on unabsorbed carbohydrates, or for carbon dioxide, which is derived from metabolism of an orally administered isotopically labeled test substance, and can be used to detect bilary, pancreatic, and mucosal cell malfunction as well as bacterial overgrowth in the small intestine. For example, the bile salt glycocholic acid is normally absorbed intact from the ileum and re-excreted in the bile. However, in patients with impaired ileal function or with bacterial overgrowth in the small intestine, there is increased bacterial deconjugation of glycocholic acid with release of glycine. Glycine is then metabolized to CO_2, primarily by bacterial enzymes. Consequently, the administration of glycocholic acid, labeled in the glycine moiety with ^{14}C, results in an increased excretion of $^{14}CO_2$ in patients with ileal disease or bacterial overgrowth. Generally, clinical validation of CO_2 breath tests has been carried out using ^{14}C radioisotopes. However, use of the stable ^{13}C analogs with quantitation of $^{13}CO_2$ by mass spectroscopy also has been employed [23].

The technique of administering an oral load and of sampling body fluid not only has clinical diagnostic importance but is a useful approach for determining the overall rate and extent of absorption of environmental contaminants. Such determinations may be important in the theoretical prediction of systemic concentrations of toxic substances following various ingestion rates. Analyses of this sort, referred to by the recently coined term *toxicokinetics*, apply mathematical tools extensively used to describe the disposition of pharmacologic agents.

A more direct approach for analyzing absorption characteristics than the sampling of systemic or excreted body fluids entails the sampling of portal blood [24] or the collection of the mesenteric blood draining the sites of absorption of the test substance. Such a procedure requires considerably more complicated surgical techniques than that of sampling systemic blood, urine, or breath. Furthermore, transfusions of blood into the animal may be required. An important advantage of this procedure is the capacity to determine *in vivo* the kinetics of metabolism of a test substance by intestinal tissues. In certain studies, the appearance of the test substance in lymph may be critical, as in the absorption of fats. Cannulation of the mesenteric lymphatic vessel may be carried out even in a small animal such as the rat [25].

Another approach to quantitating absorption entails monitoring the disappearance of a test substance from the intestine after its administration into the lumen. The perfusion method, for example, monitors differences in the amount infused from the amount appearing at a site distal to the area of infusion. Such perfusion techniques have been exceedingly useful in determining the transport of electrolytes and nutrients in man. The determination of the amount unabsorbed in the sample taken at a distal site is made possible by the use of a marker substance that is neither metabolized nor absorbed, commonly polyethylene glycol 4000. This particular marker has the advantages of lack of adsorption to the gut, high water solubility, stability during frozen storage, and ease of determination by radioisotopic or spectrophotometric methods [26]. However, one drawback to techniques in which only the luminal contents are sampled is that retention of the test substances in the intestinal mucosa is not quantitated.

In the closed segment procedure, referred to above, the extent of absorption is calculated on the basis of disappearance of the test substance from both the lumen and the intestinal tissue. With the closed segment method, at the end of the absorption period, the entire segment, both intestinal wall and contents, is assayed quantitatively for the amount of the test substance remaining. A disadvantage to this technique when compared with perfusion methods is that sequential samples cannot be taken from a single animal. However, this method is readily used in small animals and can therefore be relatively economical.

In both the perfusion and the closed segment procedures, equating loss of a test substance with its absorption requires verification that disappearance does not result as a consequence of metabolism in the intestine. If metabolism of the substance does occur, then assay of the intestine alone is inadequate for a description of its absorption kinetics, unless the metabolite is poorly absorbed and can be completely recovered in the intestinal samples.

Another important safety assay of the gastrointestinal system is the influence of test article on the formation of ulcers [27]. After overnight fasting, young rats are given the test article and euthanized 4 or 6 h later. The mucosal surface of the stomach and duodenum is scored for the presence of hyperemia, hemorrhage, and ulcers. The dose-dependent ulcerative properties of NSAIDs are clearly demonstrated in this assay, making it important in the development of other NSAIDs that are not as caustic to the gastrointestinal mucosa [28].

Additional digestive system safety pharmacology tests include effects of test articles on gastric emptying rate and gastric secretion. Gastric emptying rate is measured in rats using a solution of phenol red (or Evans blue) delivered via oral gavage a preset time after administration of the test article [29]. The dilution of phenol red after 30 min in the rat's stomach is determined colorimetrically at 558 nm in a spectrophotometer. This is compared with a group of control rats that are euthanized immediately after phenol red administration. The influence of test articles on gastric juice secretion is accomplished by ligating the pyloric sphincter under anesthesia in rats following a fasting period [11,30]. Immediately after recovery from anesthesia, each rat is given a preset dose of the test article. The fluid content of the rat's stomach is recovered after a set period of time, usually 4 h. The volume and contents of the stomach are measured to determine the effect of the test article on gastric secretions. Electrolyte concentrations, pH, and protein content of gastric secretions can be measured in this assay [30].

8.5   Gastric emptying rate and gastric pH changes: A new model

Sometimes new technologies for safety pharmacology can come from clinical settings. The Heidelberg pH Capsule (HC) was developed over 30 years ago at Heidelberg University in then West Germany. H.G. Noller invented and first tested this device on more than 10,000 adult patients over a 3-year period. The HC is a pill-sized device containing an antimony-silver chloride electrode for measuring pH and a high-frequency transmitter operating at an average frequency of 1.9 MHz. The transmitter in the HC is activated by immersion in physiologic saline by a permeable membrane enclosing the battery compartment. Thus, when a patient swallows the HC, the fluid contents of the stomach activate the transmitter. Transmitted signals are picked up via a belt receiver and can be displayed and recorded. The profile of changes in pH over time correlate with the movement of the HC through the different regions of the gastrointestinal tract [31]. The pH of the fasted human stomach is very acidic, on average about pH 1. When the HC moves through the pyloric sphincter and into the duodenum, there is a rapid increase in pH of more than 4 pH units. Thus, one can get a fairly precise measure of gastric emptying rate in humans with this noninvasive technique. Additional pH changes have been correlated with transition of the HC through the duodenum, jejunum, and colon.

Mojaverian and colleagues have used the HC extensively to examine the influence of gender, posture, age, and content and frequency of food ingestion on the gastric emptying rate (or gastric residence time) in healthy volunteers [31,32]. Although developed for clinical use in people, the HC may be a useful tool for measuring important digestive system parameters in laboratory animals. The size of the HC, approximately the size of a No. 1 gelatin capsule (7 mm diameter, 20 mm long) prohibits its use in small animals [31]. It may be useful in studies with dogs and possibly in nonhuman primates. In particular, the HC could be used to measure gastric emptying rate in a totally noninvasive manner in dogs [33]. Dogs are readily trainable to accept pills and to wear a receiver belt and could be tested after administration of a test compound [34, 35]. This technique for measuring gastric emptying rate in dogs is also advantageous in that it is not a terminal procedure. The influence of test articles on the pH within different portions of the gastrointestinal system could also be measured with the HC [36]. The major drawback for using the HC for safety pharmacology screening is the price of the capsules and the receiver system.

8.6 Effects of drugs on gut immune system (jejunum, ileum, colon)

- Histological damage score
  - Macroscopic
  - Microscopic
- Resident and attracted immunocytes
  - Mast cell numbers / degranulation *in vitro*
  - Neutrophil activation
    - Tissue myeloperoxidase (MPO)
    - Fecal calprotectin
  - Lymphocyte infiltration
- Cytokine profile
  - TH1 (IL2, IFNγ) / Th2 (IL4, IL5)
  - IL1β, TNFα

8.7 Candidate drugs to evaluate for effects on gut immune system

- PDE inhibitors (PDE III and IV):
  - Rolipram:
    - Gastric glandular mucosa inflammation
    - Reduced neutrophilic attraction (colon)
    - Intestinal edema
    - Increased mucus secretion
- Other candidates:
  - Kinase inhibitors (Rho and MAP kinase)
  - Herbal preparations
  - NO (nitrous oxide) donors

Specific methods also exist for using the rat to evaluate the potential of drugs to cause or influence gastrointestinal inflammation and ulceration [28].

8.8   Conclusions

1. Gastrointestinal side effects of drugs are frequent and not limited to alterations in GI transit or motility or histological damage observed in toxicology studies.
2. Influence of drugs on mucosal barrier and particularly on paracellular permeability may have long-term effects on gut immune system and immune equilibrium with colonic microflora (food allergy or intolerance, leakage of immunocytes, macromolecules, bacterial infiltrations).
3. Direct or indirect effects of drugs on sensitivity may trigger chronic symptoms similar to FBD and particularly IBS affecting the quality of life.

References

1. Canadian Pharmaceutical Association, *Compendium of Pharmaceuticals and Specialties*, 30th ed., Ottawa, Ontario, Canada.
2. Gad, S.C., *Drug Safety Evaluation*, 2nd ed., New York: John Wiley & Sons, 2009.
3. Gad, S.C., *Target Organ Toxicity: The GI Tract*, 2nd ed., Philadelphia: Taylor & Francis, 2007.
4. Bueno, L., *Gastrointestinal Safety Pharmacology: Exploring More Than Just Gastro-Intestinal Motility*, Lyon, France: MDS Pharma Safety Pharmacology Symposium, December 2002.
5. Mortin, L.I., Horvath, C.J., and Wyand, M.S., Safety pharmacology screening: practical problems in drug development, *Int. J. Toxicol.*, 1997, 16:41–65.
6. Bjarnason, T., et al., Importance of local versus systemic effects of non-steroidal ant-inflammatory drugs in increasing small intestinal permeability in man, *Gut*, 1991, 32:275–277.
7. Lacroix, P., and Guillaume, P., *Current Protocols in Pharmacology*, 5.3.1–5.3.8, Gastrointestinal models: intestinal transit and ulcerogencic activity in the rat, 1998.
8. Gad, S.C. (ed.), *Animals Models in Toxicology*, 2nd ed., Philadelphia: Taylor & Francis, 2006.
9. Lavigne, M.E., et al., Gastric emptying rates of solid food in relation to body size, *Gastroenterology*, 1978, 74:1258–1260.
10. Hinton, J.M., Lennard-Jones, J.E., and Young, A.C., A new method for studying gut transit times using radio-opaque markers, *Gut*, 1969, 10:842–847.
11. Summers, R.W., Kent, T.H., and Osborne, J.W., Effects of drugs, ileal obstruction, and irradiation on rat gastrointestinal propulsion, *Gastroenterology*, 1970, 59:731–739.
12. Hightower, N.C., Motor action of the small bowel. In: *Handbook of Physiology, Sec. 6, Alimentary Canal, Vol IV, Motility* (Code, C.F., ed.), Washington, D.C.: American Physiological Society, 1968, pp. 2001–2024.
13. Scott, L.D., and Summers, R.W., Correlation of contractions and transit in rat small intestine, *Am. J. Physiol.*, 1976, 230:132–137.

14. Bass, P., *In vivo* electrical activity of the small bowel. In: *Handbook of Physiology, Sect. 6: Alimentary Canal, Vol. IV: Motility* (Code, C.F. ed.), Washington, D.C.: American Physiological Society, 1968, pp. 2051–2076.

15. Scultz, S.G., Frizzell, R.A., and Mellans, H.M., Ion transport by mammalian small intestine, *Ann. Rev. Physiol.*, 1974, 36:51–91.

16. Parsons, D.S., Methods for investigation of intestinal absorption. In: *Handbook of Physiology, Section 6, Alimentary Canal, Vol. III, Intestinal Absorption* (Code, C.F., ed.), Washington, D.C.: American Physiological Society, 1968, pp. 1177–1216.

17. Levine, R.R., Intestinal absorption. In: *Absorption phenomena* (Rabinowitz, J.L., and Myerson, R.M., eds.), New York: Wiley-Interscience, 1971, pp. 27–96.

18. Fordtran, J.S., et al., Permeability characteristics of the human small intestine, *J. Clin. Invest.*, 1965, 44:1935–1944.

19. Thomas, A.B.R., and Dietschy, J.M., Intestinal absorption: major extracellular and intracellular events. In: *Physiology of the gastrointestinal Tract, Vol. 2*, New York: Raven Press, 1981, pp. 1147–1220.

20. Levine, M.E., and Pelikan, E.W., The influence of experimental procedures and dose on the intestinal absorption of an onium compound, benzomethamine. *J. Pharmacol. Exp. Ther.*, 1961, 131:319–327.

21. Wagner, J.G., *Fundamentals of Clinical Pharmacokinetics*. Hamilton, IL: Drug Intelligence Publications, Inc., 1975, p. 173.

22. Gray, G., Maldigestion and malabsorption-clinical manifestations and specific diagnosis. In: *Gastrointestinal Disease: Pathophysiology, Diagnosis, Management* (Sleisenger, M.H., and Fordtran, J.S., eds.), Philadelphia: W.B. Saunders, 1978, pp. 36–78.

23. Schwabe, A.D., and Hepner, G.W., Breath tests for the detection of fat malabsorption, *Gastroenterology*, 1979, 76:216–218.

24. Watkins, J.B., et al., C-trioctanoin: a nonradioactive breath test to detact fat malabsorption, *J. Lab. Clin. Med.*, 1977, 90:422–430.

25. Pelzmann, K.S., and Havemeyer, R.N., Portal vein blood sampling in intestinal drug absorption studies. *J. Pharm. Sci.*, 1971, 60:331.

26. DeMarco, T.J., and Levine, R.R., Role of the lymphatics in the intestinal absorption and distribution of drugs, *J. Pharmacol. Exp. Ther.*, 1969, 169:142–151.

27. Soergel, K.H., Inert markers, *Gastroenterology*, 1968, 54:449–452.

28. Shay, H., et al., A simple method for the uniform production of gastric ulceration in the rat, *Gastroenterology*, 2005, 5:43–61.

29. Whitely, P.E., and Dabrymple, S.A., *Current Protocols in Pharmacology*, 10.2.1–10.2.4, Models of Inflammation: Measure Gastrointestinal Ulceration in the Rat, 1998.

30. Megens, A.A.H.P., Awouters, F.H.L., and Niemegeers, C.J.E., General pharmacology of the four gastrointestinal motility stimulants bethanechol, metoclopramide, trimebutine, and cisapride, *Arzneim.-Forsch./Drug Res.*, 1991, 41(I):631–634.

31. Tamhane, M.D., et al., Effect of oral administration of *Terminalia chebula* on gastrix emptying: an experimental study, *J. Postgrad. Med.*, 1997, 43(I):12–13.

32. Mojaverian, P., et al., Gastrointestinal transit of a solid indigestible capsule as measured by radiotelemetry and dual gamma scintigraphy, *Pharm. Res.*, 1989 6:717–722.

33. Mojaverian, P., et al., Mechanism of gastric emptying of a nondisintegrating radiotelemetry capsule in man, 1991, *Pharm. Res.* 8:97–100.
34. Itoh, T., et al., Effect of particle size and food on gastric residence time of non-disintegrating solids in beagle dogs, *J. Pharm. Pharmacol.*, 1986, 38:801–806.
35. Lui, C.Y., et al., Comparison of gastrointestinal pH in dogs and humans: implications on the use of the beagle dog as a model for oral absorption in humans, *J. Pharm. Sci.*, 1986, 75:271–274.
36. Vashi, V.I., and Meyer, M.C., Effect of pH on the *in vitro* dissolution and *in vivo* absorption of controlled-release theophylline in dogs, *J. Pharm. Sci.* 1988, 77:760–764.
37. Youngberg, C.A., et al., Radio-telemetric determination of gastrointestinal pH in four healthy beagles, *Amer. J. Vet. Res.*, 1985, 46:1516–1521.

chapter nine

The immune system

9.1 Introduction to the immune system and adverse modulation activities of drugs

At the time of the introduction of the International Conference on Harmonisation of Technical Requirements for Registration of Pharmaceuticals for Human Use (ICH) safety pharmacology guidance (S7 A&B), there was significant dissent over the "pivotal" organ systems that were required to be assessed before human exposure to a new drug, not including the immune system. Like the three organ systems that are listed in S7 as primary (cardiovascular, respiratory, and central nervous system, or CNS), transitory pharmacology effects on the immune system can also be acutely life-threatening (as demonstrated in 2006 with the CD28-Mab, TGNI-412), and its exclusion caused some dismay.

Exclusion from such listing was due to the planned implementation of a specific guidance on the evaluation of immunotoxicology of new drugs—ICH S8 [1], not implemented, however, until 2005.

The immune system was (and still is) listed as a secondary organ system for safety pharmacology evaluation in S7A [2]. In this context, it is presented here.

The immune system is a highly complex system of cells involved in a multitude of functions including antigen presentation and recognition, amplification, and cell proliferation with subsequent differentiation and secretion of lymphokines and antibodies. The end result is an integrated system responsible for defense against foreign pathogens and spontaneously occurring neoplasms that, if left unchecked, may result in infection and/or malignancy. To be effective, the immune system must be able to both recognize and destroy foreign antigens. To accomplish this, cellular and soluble components of diverse function and specificity circulate through blood and lymphatic vessels, thus allowing them to act at remote sites and tissues. For this system to function in balance and harmony requires regulation through cell-to-cell communication and precise recognition of self versus nonself, with appropriate recognition and response to potential damaging or dangerous challenges. Immune active agents can upset this balance if they are lethal to one or more of the cell types or alter membrane morphology and receptors. Several undesired immune system responses may occur upon repeated therapeutic administration of a pharmaceutical that may ultimately present barriers to its development, including the following:

- Down-modulation of the immune response (immunosuppression), which may result in an impaired ability to deal with neoplasia and infections. This is of particular concern if the therapeutic agent is intended to be used in patients with pre-existing conditions such as cancer, severe infection, or immunodeficiency diseases.
- Up-modulation of the immune system (i.e., autoimmunity).
- Direct adverse immune responses to the agent itself in the form of hypersensitivity responses (e.g., anaphylaxis and delayed contact hypersensitivity).
- Direct immune responses to the agent that limit or nullify its efficacy (i.e., the development of neutralizing antibodies).

The safety pharmacology evaluation of potential drugs, unlike its close relative immune toxicology, is limited in scope and not part of the required testing before a new drug is administered to man. This is probably because the separate immunotoxicology guidelines cover much ground and an overlap is undesirable. The initial (and only) required evaluation of a drug in clinical development was previously limited to an evaluation of potential to induce passive cutaneous anaphylaxis (PCA). Immune modulated responses to drugs (drug allergies) are a major problem and cause of discontinuance of use by patients who do need access to the therapeutic benefits [3], and there remains no adequate preclinical methodology for identifying or predicting these responses to orally administered small molecule drugs [4].

As a discipline, immunopharmacology involves the study of the effects that xenobiotics have on the immune system. Several different types of adverse immunological effects may occur, including immunosuppression, autoimmunity, and hypersensitivity. Although these effects are clearly distinct, they are not mutually exclusive. For example, immunosuppressive drugs that suppress suppressor-cell activity can also induce autoimmunity [5], and drugs that are immunoenhancing at low doses may be immunotoxic at high doses. Although, in general, therapeutic agents are not endogenously produced, immunologically active biological response modifiers that naturally occur in the body should also be included, since many are not known to compromise immune function when administered in pharmacologically effective doses [6].

Although the types of immunological responses to various therapeutics may be similar, the approach taken for screening potential immunological activity will vary depending on the route of administration of the compound. Pharmaceuticals are developed with intentional but restricted human exposure, and their biological effects are extensively studied in surveillance. Pharmaceuticals are developed to be biologically active and, in some cases, intentionally immunomodulating or immunosuppressive. Many will react with biological macromolecules or require receptor binding to be pharmacologically active. By their nature, these interactions may result in altering the function of the cells of the immune system, may adversely alter the appearance of self to produce an autoimmune response, or may form a hapten, which may then elicit a hypersensitivity response. Because of the fast-expanding development of new drugs that can potentially impact the immune responsiveness of humans, immune function testing of new pharmaceutical products has become a growing concern.

Until recently, immune function evaluation in pharmaceutical safety assessment has been poorly addressed by both regulatory requirements and guidelines and by existing practice. Notable exceptions are the testing requirements for delayed contact hypersensitivity for dermally administered agents and antigenicity/anaphylaxis testing for drugs to be registered in Japan. Unanticipated immunotoxicity is infrequently observed with drugs that have been approved for marketing. With the exception of drugs that are intended to be immunomodulatory or immunosuppressive as part of their therapeutic mode of action, there is little evidence that drugs cause unintended functional immunosuppression in man [7]. However, hypersensitivity (allergy) and autoimmunity are frequently observed and are serious consequences of some drug therapies [3,8,9,10]. An adverse immune response in the form of hypersensitivity is one of the most frequent safety causes for withdrawal of drugs that have already made it to the market (see Table 9.1) and accounts for approximately 15% of adverse reactions to xenobiotics [12]. In addition, adverse immune responses such as this (usually urticaria and frank rashes) are the chief unexpected finding in clinical studies. These findings are unexpected in that they are not predicted by preclinical studies because there is a lack of good preclinical models for predicting systemic hypersensitivity responses, especially to orally administered agents. As a consequence, the unexpected occurrence of hypersensitivity in the clinic may delay, or even preclude, further development and commercialization. Thus, a primary purpose for preclinical immunotoxicology testing is to help us detect these adverse effects earlier in development, before they are found in clinical trials.

9.1.1 Passive cutaneous anaphylaxis: Test for potential antigenicity of compound

PCA evaluation is a screen for a form of potential immune stimulation and can be performed using mice, rats, or guinea pigs (historically, primates have also been used).

Table 9.1 Drugs Withdrawn from the Market Due to Dose- and Time-Unrelated Toxicity Not Identified in Animal Experiments [11]

Compound	Adverse reaction	Year of introduction	Years on the market
Aminopyrine	Agranulocytosis	Approx 1900	75
Phenacetin	Interstitial nephritis	Approx 1900	83
Dipyrone	Agranulocytosis	Approx 1930	47
Clioquinol	Subacute myelo-optic neuropathy	Approx 1930	51
Oxyphenisatin	Chronic active hepatitis	Approx 1955	23
Nialamide	Liver damage	1959	19
Phenoxypropazine	Liver damage	1961	5
Mebanazine	Liver damage	1963	3
Ibufenac	Hepatotoxicity	1966	2
Practolol	Oculo-mucocutaneous syndrome	1970	6
Alcolofenace	Hypersensitivity	1972	7
Azaribine	Thrombosis	1975	1
Ticrynafen	Nephropathy	1979	1
Benoxaprofen	Photosensitivity, hepatotoxicity	1980	2
Zomepirac	Urticaria, anaphylactic shock	1980	3
Zirnelidine	Hepatotoxicity	1982	2
Temafloxacin	Hepato- and renal toxicity	1990	2
Tronan	Hepato- and renal toxicity	1997	3
Renzalin	Hepatotoxicity	1996	4

Anaphylaxis is an immediate-type hypersensitivity reaction involving specific IgE antibodies [13]. Following a sensitizing contact, IgE bind to high-affinity receptors on mast cells and basophils. After a subsequent contact, the reaction between a divalent antigen and bound IgE results in the degranulation of target cells with the immediate release of stored vasoactive mediators (histamine) and the synthesis of arachidonic acid derivatives (e.g., prostaglandins and leucotrienes). These mediators exert a wide array of biological effects, which account for the clinical symptoms of anaphylaxis such as urticaria, angioedema, bronchospasm, and shock.

Drug-induced anaphylaxis is a relatively rare but life-threatening event [14]. It has been reported with protein-derived drugs (e.g., heparin, insulin), penicillin, curates, and miscellaneous drugs.

The IgE-mediated PCA reaction in the rat was one of the first *in vivo* animal models in which cromolyn was shown to be effective and has since been used extensively in screening for similar compounds. In this method, inflammatory mediators released by immediate hypersensitivity reactions in the skin produce a local increase in capillary permeability.

Measurement of this response provides an estimate of the intensity of the cutaneous anaphylactic reaction [15].

Rats are generally injected intradermally with serum containing IgE antibodies to antigens such as ovalbumen. After a latent period of 24 to72 h to permit sensitization of the cutaneous mast cells, the animal is injected intravenously with a mixture of antigen and marker dye. The animal is subsequently killed and the dorsal skin deflected to reveal the undersurface. Extravasation of the dye is then estimated by determining the diameter of the odernatous lesion or by extraction and spectrophotometric determination of the marker. Alternatively, extravasation of a high-molecular-weight radioactive tracer can be employed. Test drugs are usually injected intravenously together with the antigen or, where appropriate, administered orally beforehand.

Both cromolyn and nedocromil are effective inhibitors in this system and exhibit a comparable potency. The two compounds exhibit tachyphylaxis and cross-tachyphylaxis [16] (see below), which is normally taken to imply that they are acting through a common mechanism.

At first sight, the rat PCA test has many apparent advantages for the screening of novel antiallergic compounds. It is rapid and simple, and large numbers of experimental drugs can be readily monitored. However, it depends on the ability of these compounds to inhibit degranulation of rat skin mast cells. Given the heterogeneity of mast cell responses to antiallergic agents, it is by no means clear to what extent these findings can be extrapolated to man. Certainly, many compounds have been developed that are very much more potent that cromolyn in the rat PCA test but that have no value in the management of clinical asthma.

The guinea pig has been considered to be an appropriate test species because of the histologic similarities that exist between the lungs of antigen-exposed animals and asthmatic patients and because the cavy can exhibit early- and late-onset airway obstruction, bronchial eosinophilia, and acquired bronchial hyperreactivity.

Bronchial challenge of ovalbuman-sensitized guinea pigs produces a triphasic reduction in specific airways conductance (sGaw), with maximal reductions at 2, 17, and 72 h, accompanied by infiltration of the airways with neutrophils at 17 h and eosinophils at 17 and 72 h. Nedocromil inhaled before challenge blocks the 2 and 17 h sGaw response but not the neutrophil influx, indicating that these effects are unrelated. The sGaw response and the eosinophil accumulation at 72 h are also inhibited by nedocromil given at this time.

Repeated exposure to ovalbumin aerosol produces a significant increase in the number of epithelial eosinophils in the airways of all sizes and induces bronchial hyperreactivity as assessed by an increased pulmonary resistance to inhaled acetylcholine. These changes are also inhibited by nedocromil.

9.1.1.1   Test method for pulmonary sensitization [18]

1. Shave and depilate naïve guinea pigs 24 h prior to use. One guinea pig can serve as the recipient animal for two serum samples; each serum sample must be evaluated on two animals since two antigens (PA-GPSA and GPSA) are used at challenge.
2. Using an indelible ink marker, mark seven injection sites along each shaved side of the animals; the sites should be approximately 5 cm apart.
3. Dilute the test sera 1:4, 1:8, 1:16, 1:32, 1:64, and 1:128 in physiological saline; minimal volume needed is 300 μl.
4. Using the 26-gauge needle, i.d. inject 100 μl of each dilution of serum sample at six of the marked sites along one side of the guinea pig; dilutions of a second serum sample are injected at the marked sites along the other side of the animal. The seventh marked site on each side of the animal is injected in two naïve guinea pigs.
5. Prepare the challenge antigen solutions by making 500 μl/ml solutions of PA-PGSA or GPSA in 1.0% Evans blue dye.
6. Weigh each animal and lightly ether anesthetize 4 h after the i.d. injections.
7. Using the 21-gauge needle, i.c. inject 2.0 ml/kg of the challenge antigen solutions intracutaneously into the appropriate animals (i.e., a 400-g guinea pig is i.c. injected with 0.8 ml of challenge solution.)
8. Examine the injected skin sites for signs of bluing 15–30 min after the antigen challenge. The PCA titer is defined as the reciprocal of the highest dilution of serum to yield a significant blue response at the injection site. This response must be greater than the bluing observed at the saline-injected site. A significant antibody response can be typified by solid blue coloration extending as a circle beyond the injection site or a continuous blue ring around the injection site (halo effect) or a discontinuous blue arc around the injection site. The blue response that occurs at the saline sites should not exceed more than 2–3 mm beyond the puncture site. Significant antibody to PA is defined as a PCA titer in PA-GPSA–challenged PCA animals but not in GPSA–challenged PCA animals. If an end point titer cannot be reached at the 1:128 dilution of serum, repeat the PCA test with additional dilutions of sera that range from 1:64 to 1:2048 (or higher if required).

There are several approaches to analyzing the presenting data from the PCA test. One simple approach is to express the percent PCA-positive sera in each test group. A second approach is to calculate the mean PCA titer for each dose group and use a simple statistical package for determining significant differences in titer versus injected dose of chemical. A third approach is to convert the PCA dilution titers into the logarithm to the base 2 of the reciprocal of the end point dilution (i.e., 1:4 dilution equals a titer of 2 and 1:1024 dilution equals a titer of 10) and determine the significant differences from the log base 2 values.

The general pitfalls in the PCA test are comparable to the pitfalls encountered in the ACA (active cutaneous anaphylaxis) test. There is the risk of anesthesia and the potential loss of animals from the i.c. injection. In addition, on occasion, the PCA-recipient animals can be dermatographic and develop a mottled blue appearance on the skin. These nonspecific skin responses can interfere with evaluating the PCA response at the serum injection sites. The subjectiveness in reading the PCA titer is not as great as it is in the ACA due to the differences in the appearance of the skin responses. In the ACA test, skin reactions occur as solid blue spots that decrease in size with decreasing concentrations of antigen. In the PCA test, significant antibody at the lower dilutions yield solid blue spots the size of a dime (1.8 cm), whereas antibody at the higher dilutions yield responses that are not solid blue in the center but are fully or partially ringed. Therefore, it is easier to distinguish between a true allergic response in the skin versus irritation due to saline injections in the PCA test as compared with the ACA test.

9.1.2   Center for Drug Evaluation and Research guidance for investigational new drugs

The Center for Drug Evaluation and Research (CDER) recently (2001) promulgated draft guidance for pre-IND immunotoxicity [19]; it has been open for comment and is certain to be modified some, but it clearly establishes the framework for Food and Drug Administration's (FDA's) approach. It begins by characterizing five adverse event categories:

- immunosuppression
- antigenicity
- hypersensitivity
- autoimmunity
- adverse immunostimulation

Specific tests are proposed for each of these categories. It notes that immune system effects in nonclinical toxicology studies are often attributed and written off as due to stress [20]. Such effects are frequently reversible with repeat dosing and tend not to be dose-related. It is also proposed that, when possible, dose extrapolations to those in clinical use be based on relative body area. Specific recommendations are made for when to conduct specific testing (as opposed to the broader general evaluations integrated into existing repeat-dose testing) (Figure 9.1) and for follow-up studies for exploring mechanisms (Figure 9.2).

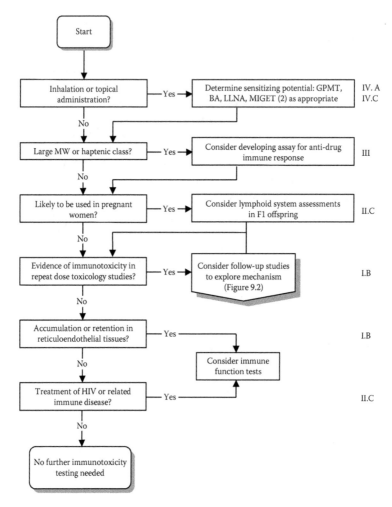

1. Annotations in right margin indicate location of text describing specific advice
2. GPMT—Guinea pig maximization test: BA— Buehler assay (Buehler patch test); LLNA—Local lymph node assay;
   MIGET—Mouse IgE test (There is only a relatively small database available for assessing the usefulness of the
   MIGET for drug regulatory purposes)

Figure 9.1 CDER flowchart for determining when to conduct specific immuno-toxicity testing [19].

9.2   Overview of the immune system

A thorough review of the immune system is not the intent of this chapter, but a brief description of the important components of the system and their interactions is necessary for an understanding of how xenobiotics can affect immune function. A breakdown at any point in this intricate and dynamic system can lead to immunopathology.

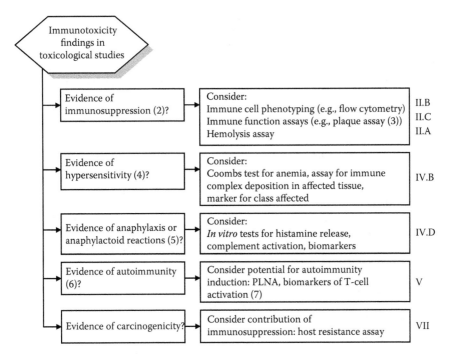

1. Annotations in right margin indicate location of text describing specific advice.

2. Examples include myelosuppression, histopathology in immune associated tissues, increased infection, tumors, decreased serum Ig, phenotypic changes in immune cells.

3. Other acceptable assays include drug effect on NK cell function *in vitro* bastogenesis, cytotoxic T-cell function, cytokine production, delayed-type hypersensitivity, host resistance to infections or implanted tumors.

4. Examples include anemia, leukopenia, thrombocytopenia, pneumonitis, vasculitis, lupus-like reactions, glomerulonephritis.

5. Examples include cardiopulmonary distress, rashes, flushed skin, swelling of face or limbs.

6. Examples include vasculitis, lupus-like reactions, glomurelonephritis, hemolytic anemia.

7. There are no established assays that reliably assess potential for autoimmunity and acute systemic hypersensitivity. The popliteal lymph node assay (PLNA) has only a relatively small database available for assessing its usefulness for drug regulatory purposes.

Figure 9.2 Follow-up studies to consider for exploring mechanisms of immuno-toxicity.

The immune system is divided into two defense mechanisms: non-specific, or innate, and specific, or adaptive, mechanisms that recognize and respond to foreign substances. Some of the important cellular components of nonspecific and specific immunity are described in Table 9.2. The nonspecific immune system is the first line of defense against infectious organisms. Its cellular components are the phagocytic cells such as the monocytes, macrophages, and polymorphic neutrophils (PMNs).

Table 9.2 Cellular Components of the Immune System and Their Functions

Cell subpopulations	Markers[a]	Functions
Nonspecific immunity		

Granulocytes
  Neutrophils (blood)
  Basophils (blood)
  Eosinophils (blood)
  Mast cells (connective tissue)
		Degranulate to release mediators

Natural killer cells (NK)		Nonsensitized lymphocytes; directly kill target cells

Reticuloendothelial
  Macrophage
    (peritoneal, pleural,
    alveolar spaces)
  Histiocytes (tissues)
  Monocytes (blood)
	CD14; HLA-DR	Antigen processing, presentation, and phagocytosis (humoral and some cell-mediated responses)

Specific immunity		
Humoral immunity Activated B cells	CD19; CD23	Proliferate; form plasma cells
Plasma cells		Secrete antibody; terminally differentiated
Resting		Secrete IgM antibodies (primary response)
Memory		Secrete IgG antibodies (secondary response)
Cell-mediated immunity		
T-cell types: Helper (T_h)	CD4; CD25	Assists in humoral immunity; required for antibody production
Cytotoxic (T_k)	CD8; CD25	Targets lysis
Suppressor (T_s)	CD8; CD25	Suppresses/regulates humoral and cell-mediated responses

[a] Activation surface markers detected by specific monoclonal antibodies; can be assayed with flow cytometry.

The specific, or adaptive, immune system is characterized by memory, specificity, and the ability to distinguish self from nonself. The important cells of the adaptive immune system are the lymphocytes and antigen-presenting cells that are part of nonspecific immunity. The lymphocytes, which originate from pluripotent stem cells located in the hematopoietic tissues of the liver (fetal) and bone marrow, are composed of two general cell types: T and B cells. The T cells differentiate in the thymus and are

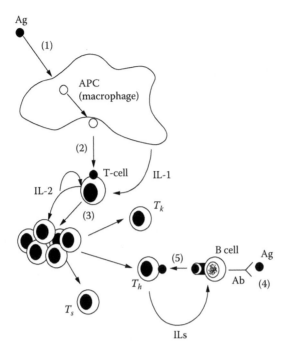

Figure 9.3 A simplified schematic of the immunoregulatory circuit that regulates the activation of T cells and B cells involved in humoral (T-cell dependent) and cell-mediated immunity. (1) Antigen (Ag) is processed by the APCs expressing class II MHC molecules. (2) Antigen plus class II MHC is then presented to antigen-specific T helper cells (CD4+), which stimulates secretion of IL-2. (3) IL-2 in turn stimulates proliferation (clonal expansion) of T cells and differentiation into T suppressor (Ts), T killer (Tk), and T helper (Th) effector cells. The expanded clone has a higher likelihood of finding the appropriate B cell that has the same antigen and class II molecules on its surface. (4) Next, the antigen binds to an antibody (Ab) on the surface of a specific B cell. (5) The B cell, in turn, processes the antigen and presents it (plus class II MHC) to the specific Th cell. The Th cell is then stimulated to secrete additional interleukins (ILs) that stimulate clonal expansion and differentiation of the antigen-specific B cell.

made up of three subsets: helper, suppressor, and cytotoxic. The B cells, which have the capacity to produce antibodies, differentiate in the bone marrow or fetal liver. The various functions of the T cells include presenting antigen to B cells, helping B cells to make antibody, killing virally infected cells, regulating the level of the immune response, and stimulating cytotoxic activity of other cells such as macrophages [21].

Activation of the immune system is thought to occur when antigen-presenting cells (APCs) such as macrophages and dendritic cells take up antigen via F_c or complement receptors, process the antigen, and present it to T cells (see Figure 9.3). Macrophages release soluble mediators such

Table 9.3 Antibodies Involved in the Humoral Immune Response [17]

Antibodies	Serum concentration Mg ml⁻¹ (%)	Characteristics/functions
IgG	10–12 (80%)	Monomeric structure (γ-globulin); secreted from B cells during secondary response; binds complement; can cross placenta
IgM	1–2 (5%–10%)	Pentameric structure; secreted from B cells during primary response; potent binder of complement; high levels indicative of systemic lupus erythematosus or rheumatoid arthritis; cannot cross placenta
IgA	3–4 (10%–15%)	Dimeric or monomeric structures; found in seromucous secretions (breast milk); secreted by B cells associated with epithelial cells in GI tract, lung, etc.
IgD	0.03 (< 1%)	Monomer; extremely labile; functions not well known
IgE	< 0.0001	Reaginic antibody involved in immediate hypersensitivity; antihelminthic; does not bind complement

as interleukin 1 (IL-1), which stimulate T cells to proliferate. APCs must present antigen to T cells in conjunction with the class II major histocompatibility complex (MHC) proteins that are located on the surfaces of T cells. The receptor on the T cell is a complex of the T1 molecule that binds antigen, the MHC proteins, and the T3 molecular complex, which is often referred to as the CD3 complex. Upon stimulation, T cells proliferate, differentiate, and express interleukin-2 (IL-2) receptors. T cells also produce and secrete IL-2, which, in turn, acts on antigen-specific B cells, causing them to proliferate and differentiate into antibody-forming (plasma) cells.

Antibodies circulate freely in the blood or lymph and are important in neutralizing foreign antigens. The various types of antibodies involved in humoral immunity and their functions are described in Table 9.3. Multiple genes (polymorphisms) encode diversity to the variable region of the antibody. B cells are capable of generating further diversity to antibody specificity by a sequence of molecular events involving somatic mutations, chromosomal rearrangements during mitosis, and recombination of gene segments [22].

The immune system is regulated in part by feedback inhibition involving complex interactions between the various growth and differentiation factors listed in Table 9.4. Since antigen initiates the signal for the immune response, elimination of antigen will decrease further stimulation [21]. T suppressor cells (T_s) also regulate the immune response and are thought to

Table 9.4 Growth and Differentiation Factors of the Immune System [23]

Factors	Cell of origin	Primary immune functions
Interleukins[a]		
IL-1	Macrophage, B and T cells	Lymphocyte-activating factor; enhances activation of T and B cells, NK cells, and macrophages
IL-2	T cells (T_h)	T-cell growth factor; stimulates T-cell growth and effector differentiation; stimulates B-cell proliferation/ differentiation
IL-3	T cells (T_h)	Mast-cell growth factor; stimulates proliferation/ differentiation of mast cells, neutrophils, and macrophages
IL-4	T cells (T_h), mast cells, B cells	B-cell growth factor; induces proliferation/differentiation of B cells and secretion of IgA, IgG_1, and IgE; promotes T-cell growth; activates macrophages
IL-5	T cells (T_h)	Stimulates antibody secretion (IgA), proliferation of B cells, and eosinophil differentiation
IL-6	T cells, fibroblasts, monocytes	Stimulates growth/differentiation of B cells and secretion of IgG; promotes IL-2-induced growth of T cells
IL-7	Bone marrow stromal cells	Stimulates pre-B- and pre-T-cell growth/differentiation; enhances thymocyte proliferation
IL-8	Monocytes, fibroblasts	Neutrophil chemotaxis
IL-9	T cells	Stimulates T cells and mast cells
IL-10	T cells	Stimulates mast cells and thymocytes; induction of class II MHC
Interferons (INF)		
A-INF	Leukocytes and mast cells	Antiviral; increases NK-cell function, B-cell differentiation, potentiates macrophage production of IL-1
B-INF	Fibroblasts, epithelial cells	Antiviral; potentiates macrophage production of IL-1; increases NK-cell function
Γ-INF	T cells (T_h), cytotoxic T cells	Antiviral; activates macrophages; induces MHC class II expression on macrophages, epithelial, and endothelial cells

Table 9.4 (*Continued*) Growth and Differentiation Factors
of the Immune System [23]

Factors	Cell of origin	Primary immune functions
Tumor necrosis factors (TNF)		
TNFα	Macrophage, B and T cells	Catectin; promotes tumor cytotoxicity; activates macrophages and neutrophils; enhances IL-2 receptor expression on T cells; inhibits antibody secretion
TNFβ	T cells (T_h)	Lymphotoxin; promotes T-cell-mediates cytotoxicity
	NK cells	B-cell activation
Colony stimulating factors (CSF)		
	Stem cells:	*Promotes growth and differentiation of:*
Granulocyte CSF	Myeloid	Granulocytes and macrophages
Macrophage CSF	Myeloid	Macrophages and granulocytes
Granulocyte-macrophage CSF	Myeloid	Granulocytes, macrophages, eosinophils, mast cells, and pluripotent progenitor cells

a Includes lymphokines, monokines, and cytokines produced by T cells, macrophages, and other cells, respectively.

be important in the development of tolerance to self antigens. In addition to the humoral immune system or the branch that is modulated by antibody, cell-mediated immunity and cytotoxic cell types play a major role in the defense against virally infected cells, tumor cells, and cells of foreign tissue transplants. Cytotoxic T_k cells (T killer cells) recognize antigen in association with class I molecules of the MHC, while natural killer cells (NK cells) are not MHC restricted. Cell killing results in a sequence of events following activation of the effector cell, lysosomal degranulation, and calcium influx into the targeted cell. The various types of cells involved in cell-mediated cytotoxicity and their mechanisms of action are outlined in Table 9.5.

9.3　Immunotoxic effects

The immune system is a highly integrated and regulated network of cell types that requires continual renewal to achieve balance and immunocompetence. The delicacy of this balance makes the immune system a natural target for cytotoxic drugs or their metabolites. Since renewal is dependent on the ability of cells to proliferate and differentiate, exposure to agents that arrest cell division can subsequently lead to reduced immune function or immunosuppression. This concept has been exploited in the development of therapeutic drugs intended to treat leukemias, autoimmune

Table 9.5 Cells and Mechanisms Involved in Cell-Mediated Cytotoxicity

Cell type	Mechanism of cytotoxicity
T_k cells	T_k cells that are specifically sensitized to antigens on target cells interact directly with target cells to lyse them.
T_D	Cells involved in delayed hypersensitivity that act indirectly to kill target cells; T_D cells react with antigen and release cytokines that can kill target cells.
NK cells	Nonspecific T cells that react directly with target cells (tumor cells) without prior sensitization.
Null cells	Antibody-dependent cell-mediated cytotoxicity (ADCC) involving non-T/non-B cells (null cells) with F_c receptors specific for antibody-coated target cells.
Macrophages	Nonspecific, direct killing of target by phagocytosis; also involved in presenting antigen to specific T_k cells that can then mediate cytotoxicity as described above.

disease, and chronic inflammatory diseases and to prevent transplant rejection. However, some drugs may adversely modulate the immune system secondarily to their therapeutic effects.

Two broad categories of immunotoxicity have been defined on the basis of suppression or stimulation of normal immune function. Immunosuppression is a down-modulation of the immune system characterized by cell depletion, dysfunction, or dysregulation that may subsequently result in increased susceptibility to infection and tumors. In contrast, immunostimulation is an increased or exaggerated immune responsiveness that may be apparent in the form of a tissue-damaging allergic hypersensitivity response or pathological auto-immunity. However, as knowledge of the mechanisms involved in each of these conditions has expanded, the distinction between them has become less clear. Some agents can cause immunosuppression at one dose or duration of exposure, and immunostimulation at others. For instance, the chemotherapeutic drug cyclophosphamide is in most cases immunosuppressive; however, it can also induce autoimmunity [5]. Likewise, dimethylnitrosamine, a nitrosamine detected in some foods, has been shown to have both suppressing and enhancing effects on the immune system [24].

9.3.1 Immunosuppression

The various cells of the immune system may differ in their sensitivity to a given xenobiotic. Thus, immunosuppression may be expressed as varying degrees of reduced activity of a single cell type of multiple populations of immunocytes. Several lymphoid organs such as the bone marrow, spleen, thymus, and lymph nodes may be affected simultaneously or the immunodeficiency may be isolated to a single tissue, such as the Peyer's patches of the intestines. The resulting deficiency may in turn lead to an array of clinical outcomes of varying ranges of severity. These outcomes include increased susceptibility to infections, increased severity or persistence of infections, or infections with unusual organisms (e.g., systemic fungal infections). Immunosuppression can be induced in a dose-related manner by a variety of therapeutic agents at dose levels lower than those required to produce overt clinical signs of general toxicity. In addition, immunosuppression can occur without regard to genetic predisposition, given that a sufficient dose level and duration of exposure has been achieved.

Humoral immunity is characterized by the production of antigen-specific antibodies that enhance phagocytosis and destruction of microorganisms through opsonization. Thus, deficiencies of humoral immunity (B lymphocytes) may lead to reduced antibody titers and are typically associated with acute gram-positive bacterial infections (i.e., *Streptococcus*). Although chronic infection is usually associated with dysfunction of some aspect of cellular immunity, chronic infections can also occur when facultative intracellular organisms such as *Listeria* or *Mycobacterium* evade antibodies and multiply within phagocytic cells.

Since cellular immunity results in the release of chemotactic lymphocytes that in turn enhance phagocytosis, a deficiency in cellular immunity may also result in chronic infections. Cellular immunity is mediated by T cells, macrophages, and NK cells involved in complex compensatory networks and secondary changes. Immunosuppressive agents may act directly by lethality to T cells, or indirectly by blocking mitosis, lymphokine synthesis, lymphokine release, or membrane receptors to lymphokines. In addition, cellular immunity is involved in the production and release of interferon, a lymphokine that ultimately results in blockage of viral replication (Table 9.2). Viruses are particularly susceptible to cytolysis by T cells since they often attach to the surface of infected cells. Thus, immunosuppression of any of the components of cellular immunity may result in an increase in protozoan, fungal, and viral infections as well as opportunistic bacterial infections.

Immune system suppression may result unintentionally as a side effect of cancer chemotherapy or intentionally from therapeutics administered to prevent graft rejection. In fact, both transplant patients administered immunosuppressive drugs and cancer patients treated with chemotherapeutic agents have been shown to be at high risk of developing secondary cancers, particularly of lymphoreticular etiology [25]. Most of these drugs are alkylating or cross-linking agents that by their chemical nature are electrophilic and highly reactive with nucleophilic macromolecules (protein and nucleic acids). Nucleophilic sites are quite ubiquitous and include amino, hydroxyl, mercapto, and histidine functional groups. Thus, immunotoxic agents used in chemotherapy may induce secondary tumors through direct genotoxic mechanisms (i.e., DNA alkylation).

Reduced adaptive cellular immunity may result in increased malignancy and decreased viral resistance through indirect mechanisms as well, by modulating immune surveillance of aberrant cells. T lymphocytes, macrophage cells, and NK cells are all involved in immunosurveillance through cytolysis of virally infected cells or tumor cells, each by a different mechanism (Table 9.2) [26]. In addition to the common cell types described in Table 9.2, at least two other types of cytotoxic effector cells of T-cell origin have been identified, each of which has a unique lytic specificity phenotype and activity profile [27]. Of these, both LAK and TIL cells have been shown to lyse a variety of different tumor cells. However, TIL cells have 50–100 times more lytic activity than LAK cells. Most tumor cells express unique surface antigens that render them different from normal cells. Once detected as foreign, they are presented to the T helper cells in association with MHC molecules to form an antigen-MHC complex. This association elicits a genetic component to the immunospecificity reaction. T helper cells subsequently direct the antigen complex toward the cytotoxic T lymphocytes, which possess receptors for antigen-MHC complexes. These cells can then proliferate, respond to specific viral antigens or antigens on the membranes of tumor cells, and destroy them [24].

In contrast, the macrophages and natural killer (NK) cells are involved in nonspecific immunosurveillance in that they do not require prior sensitization with a foreign antigen as a prerequisite for lysis and are not involved with MHC molecules. The enhancement of either NK cell function or macrophage function has been shown to reduce metastasis of some types of tumors. Macrophage cells accumulate at the tumor site and have been shown to lyse a variety of transformed tumor cells [28]. Natural killer cells are involved in the lysis of primary autochthonous tumor cells. Migration of NK cells to tumor sites has been well documented. Although not clearly defined, it appears that they can recognize certain proteinaceous structures on tumor cells and lyse them with cytolysin.

9.3.2 Immunosuppressive drugs

Table 9.6 lists numerous types of drugs that are immunosuppressive and describes their immunotoxic effects. Several classes of drugs that characteristically depress the immune system are further discussed below.

9.3.2.1 Antimetabolites

This class of drugs includes purine, pyrimidine, and folic acid analogs that have been successfully used to treat various carcinomas, autoimmune diseases, and dermatological disorders such as psoriasis. Because of their structural similarities to normal components of DNA and RNA synthesis, they are capable of competing with the normal macromolecules and alkylating biological nucleophiles.

Table 9.6 Immunosuppressive Drugs and Their Effects [29]

Drugs	Biological activity and indications	Immunotoxic effects
Hormones and antagonists		
Corticosteroids (prednisone)	Anti-inflammatory; systemic lupus erythematosus; leukemias; rheumatoid arthritis; breast cancer	Depresses T- and B-cell function; reduces lymphokines; alters macrophage function; increases infections
Diethylstilbestrol	Synthetic estrogen; cancer chemotherapy	Depletes or functionally impairs T cells; enhances macrophage suppressor cell; increases infections and tumorigenesis
Estradiol	Synthetic estrogen; dysmenorrhea; osteoporosis	Decreases T_h cells and IL-2 synthesis; increases T_s cell function, infections, and tumorigenesis
Antibiotics		
Cephalosporins Chloramphenol Penicillins Rifampin Tetracyclines	β-lactam antimicrobial Wide-spectrum antimicrobial β-lactam antimicrobial Macrocyclic antibiotic Antimicrobial	Granulocytopenia; cytopenia Pancytopenia, leukopenia (idiosyncratic) Granulocytopenia; cytopenia Suppresses T-cell function Decreased migration of granulocytes
Chemotherapeutics and immunomodulators		
Arabinoside (AraA and AraC)	Antimetabolites; antivirals; leukemias; lymphomas	Leukopenia; thrombocytopenia
Azathioprine	Antimetabolite; leukemia; arthritis; transplant rejection	Inhibits protein synthesis; bone marrow suppression
Busulfan	Alkylating agent; chronic granulocytic leukemia	Leukopenia; myelosuppressive; granulocytopenia
Carmutin and Lomustin (BCNU and CCNU)	Alkylating agents; Hodgkin's disease; lymphomas	Delayed hematopoietic depression; leukopenia; thrombocytopenia
Chlorambucil	Alkylating agent; leukemia; lymphomas; vasculitis	Bone marrow suppression; myelosuppressive

Table 9.6 (Continued) Immunosuppressive Drugs and Their Effects [29]

Drugs	Biological activity and indications	Immunotoxic effects
Cyclophosphamide (Cytotoxin)	Alkylating agent; cancer chemotherapy; transplant rejection; rheumatoid arthritis	Decreased T_s cells, B cells, and NK cells
Cyclosporin A	Transplant rejections	Depresses T cells; inhibits IL-2 production
Interferon	Immunomodulator; antiviral, hairy cell leukemia	Bone marrow suppression; granulocytopenia; leukopenia
Melphalan (L-PAM)	Alkylating agent; breast and ovarian cancer	Leukopenia; bone marrow suppression; granulocytopenia; pancytopenia
6-Mercaptopurine	Antimetabolite; acute leukemias; arthritis	Decreased T-cell function; bone marrow suppression
Methotrexate	Folic acid analog; cancer chemotherapy; arthritis	Inhibits proliferation; T-cell suppression; granulocytopenia; lymphocytopenia
Penostatin	Adenosine analog; T-cell leukemia	Inhibits adenosine deaminase; suppresses T and B cells
Zidovudine (AZT)	Antiviral (HIV)	Decreases T_h cells and granulocytes
Miscellaneous		
Colchicine	Antimitotic; gout; anti-inflammatory	Inhibits migration of granulocytes; leukopenia; agranulocytosis
Diphenylhydantoin (Phenytoin)	Antiepileptic	Leukocytopenia; neutrapenia
Indomethacin (Indocin)	Nonsteroidal anti-inflammatory; analgesic; antipyretic	Neutrapenia
Procainamide	Antiarrhythmic	Agranulocytosis; leukopenia (rare)
Sulfasalazine	Antimicrobial anti-inflammatory; ulcerative colitis/ inflammatory bowel diseases	Suppresses NK cells; impaired lymphocyte function

Thioguanine and mercaptopurine are purine analogs structurally similar to guanine and hypoxanthine that have been used to treat malignancies. Azathioprine, an imidazolyl derivative of mercaptopurine, has been used as an immunosuppressive therapeutic in organ transplants and to treat severe refractory rheumatoid arthritis [30] and autoimmune disorders including pemphigus vulgaris and bullous pemphigoid. These drugs act as antimetabolites to block *de novo* purine synthesis through the erroneous incorporation of thioinosinic acid into the pathway in place of inosine. The antimetabolite can bind to the inosine receptor, which in turn will inhibit the synthesis of DNA, RNA, protein synthesis, and ultimately T-cell differentiation [31]. For example, both thioguanine and mercaptopurine can act as substrates for the HGPRT enzyme to produce T-IMP (thioinosine monophosphate) and T-GMP (thioguanine monophosphate), respectively. Thioinosine monophosphate is a poor substrate for guanylyl kinase, which would normally catalyze the conversion of GMP to GDP [32]. Thus T-IMP can accumulate in the cell and inhibit several vital metabolic reactions. At high doses, these drugs can suppress the entire immune system. However, at clinical dosages, only the T-cell response is affected, without an apparent decrease in T-cell numbers [33].

Pentostatin (2'-deoxycoformycin) is an adenosine analog that is a potent inhibitor of adenosine deaminase. Pentostatin particularly useful for treating T-cell leukemia since malignant T cells have higher levels of adenosine deaminase than most cells. Similar to individuals who are genetically deficient in adenosine deaminase, treatment with pentostatin produces immunosuppression of both T and B lymphocytes, with minimal effect on other tissues. As a result, severe opportunistic infections are often associated with its clinical use.

5-Fluorouracil (5-FU), adenosine arabinoside (AraA), and cytosine arabinoside (AraC) are pyrimidine analogs to uracil, adenine, and cytosine, respectively. 5-FU is used primarily to treat cancer of the breasts and gastrointestinal tract as well as severe recalcitrant psoriasis [34]. AraC is predominantly indicated for the treatment of acute leukemia and non-Hodgkin's lymphomas. Although high-dose therapy with AraC has a good likelihood of producing complete remission, it is often accompanied by severe leukopenia, thrombocytopenia, and anemia [35]. Likewise, myelosuppression is the major toxicity associated with bolus-dose regimens of 5-FU.

9.3.2.2   *Glucocorticosteroids*

Corticosteroids are commonly used to reduce inflammation (innate immunity), treat autoimmune diseases such as systemic lupus erythematosus (SLE), and as a prophylactic measure to prevent transplant rejection. The adrenocorticosteroid prednisone is often coadministered with other immunosuppressives such as cyclosporine and azathioprine [36].

Glucocorticosteroids act pharmacologically by modulating the rate of protein synthesis. The molecule reacts with specific receptors to form a complex that crosses into the nucleus of the cell and regulates transcription of specific mRNA. The corticosteroid complex releases inhibition of transcription, thus enhancing protein synthesis [37]. This may lead to the initiation of *de novo* synthesis of the phospholipase A2 inhibiting protein, lipocortin, which blocks the synthesis of arachidonic acid and its prostaglandin and leukotriene metabolites [38]. Glucocorticosteroids induce immunosuppression and anti-inflammation as a result of the inhibition of specific leukocyte functions such as lymphokine activity. Glucocorticoids can also inhibit recruitment of leukocytes and macrophages into the site of inflammation. In addition, amplification of cell-mediated immunity can be suppressed by inhibiting the interaction of IL-2 with its T-cell receptors. However, the immunosuppression is reversible and immune function recovers once therapy has ceased.

9.3.2.3 Cyclosporine

Cyclosporin A (cyclosporine) is an 11 amino acid cyclic peptide residue of fungal origin isolated from the fermentation products of *Trichoderma polysporum* and *Cylindrocarpon lucidum*. In addition to having a very narrow range of antibiotic activity, it was also found to inhibit proliferation of lymphocytes, which made it unsuitable as an antibiotic. Cyclosporine inhibits the early cellular response of helper T cells to antigens [39] primarily by inhibiting production of IL-2 [40], and at higher doses it may inhibit expression of IL-2 receptors [41]. Cyclosporine does not prevent the stimulation of helper T-cell clonal expansion by IL-2, but only its activation. Since it is not myelosuppressive at therapeutic dosages, the incidence of secondary infection is lower than that induced by other classes of immunosuppressives. Thus, cyclosporine is ideal as an immunosuppressive agent to prevent transplant rejection and graft–host disease [42]. Cyclosporine has also been used as an antihelminthic and as an anti-inflammatory agent to treat rheumatoid arthritis and other autoimmune-type diseases.

9.3.2.4 Nitrogen mustards

Nitrogen mustards characteristically consist of a bis(2-chloroethly) group bonded to nitrogen. These molecules are highly reactive bifunctional alkylating agents that have been successfully used in cancer chemotherapy. Included in this group are mechlorethamine, L-phenylalanine mustard (melphalan), chlorambucil, ifosfamide, and cyclophosphamide. The cytotoxic effects of each on the bone marrow and lymphoid organs are similar; however, their pharmacokinetic and toxic profiles can vary on the basis of the substituted side group. For example, the side group may consist of a simple methyl group, as is the case of mechlorethamine, or substituted phenyl groups, in the cases of melphalan and chlorambucil.

Cyclophosphamide, which contains a cyclic phosphamide group bonded to the nitrogen mustard, is representative of this class. The parent compound itself is not active *in vitro* unless treated in conjunction with an exogenous P450 microsomal enzyme system [43] such as rat liver S9 homogenate, which metabolizes it to a highly reactive alkylating agent (4-hydroxy-cyclophosphamide). Thus, *in vivo*, cyclophosphamide is not toxic until it is metabolically activated in the liver. Cyclophosphamide has been the most widely used nitrogen mustard where it has been effective as a cancer chemotherapeutic and to treat autoimmune-type diseases including SLE, multiple sclerosis, and rheumatoid arthritis [44]. Treatment with cyclophosphamide suppresses all classes of lymphoid cells, which may result in reduced lymphocyte function as well as lymphopenia and neutropenia [45]. Thus, it has also been administered as a large single dose prior to bone marrow transplants to suppress cellular immunity and subsequently inhibit rejection [46].

9.3.2.5   Estrogens

β-estradiol [47,48] and therapeutics with estrogenic activity, such as diethylstilbestrol (DES), have also been shown to be immunosuppressive [49]. Estrogens have been shown to increase T suppressor cell activity in splenocytes, decrease numbers of T helper cells, inhibit IL-2 synthesis, and modulate production of immunoregulatory factors [16]. These effects have been particularly characterized in studies with DES, a nonsteroidal synthetic estrogen used widely in the treatment of prostate and breast cancers, as well as administered to pregnant women as a "morning after" contraceptive (in the 1950s and 1960s) [50,51]. Decreased mitogenicity of human peripheral blood lymphocytes has been observed in men treated with DES for prostate cancer and women exposed *in utero* [52,53]. In mice, thymic involution and atrophy with depletion of the cortical lymphocytes have been observed histologically. Function is also modulated, as evident by depressed mixed lymphocyte responses, mitogenicity, and T-cell release of IL-2 [54]. Dean et al. [50] speculated that DES treatment selectively depletes or functionally impairs T cells and/or the induction of suppressor macrophages, resulting in immunosuppression. Macrophage suppressor cell activity is enhanced [51] and PMN cells accumulate following bacterial challenge. Although macrophage functions of phagocytosis and tumor growth inhibition are potentiated, defects in macrophage migration and decreased bactericidal activity contribute to decreased host resistance with resulting increased susceptibility to bacterial infections.

9.3.2.6   Heavy metals

Salts of some heavy metals such as gold and platinum are used pharmacologically as immunomodulators to treat rheumatoid arthritis and as antineoplastic drugs, respectively. Most heavy metal salts inhibit mitogenicity, antibody responses, and host resistance to bacterial or viral challenge, and tumor growth. Platinum has been shown to suppress humoral immunity, lymphocyte proliferation, and macrophage function [55]. Clinically, mild to moderate myelosuppression may also be evident with transient leukopenia and thrombocytopenia.

Likewise, injectable gold salts such as gold sodium thiomalate affect a variety of immune responses in man [56]. Severe thrombocytopenia occurs in 1% of patients as a result of an immunological disturbance that accelerates the degradation of platelets. Leukopenia, agranulocytosis, and fatal aplastic anemia may also occur. Although better tolerated than parenteral preparations, the organic gold compound auranofin administered orally is also immunosuppressive. In a dog study, auranofin was shown to produce thrombocytopenia similar to that described in humans administered parenteral preparations [56,57]. Long-term toxicity studies with these compounds in dogs show evidence of immune-modulating activity, possible drug-induced immunotoxicity, and treatment-related changes in immune function (e.g., lymphocyte activation).

9.3.2.7 Antibiotics

β-lactam-containing antibiotics such as the cephalosporins may also induce significant immunosuppressive effects [58,59] in a small percentage of human patients. Adverse effects including anemia, neutropenia, thrombocytopenia, and bone marrow depression were observed in dogs administered high doses of cefonicid for 6 months [57]. A similar syndrome has been characterized in cefazedone-treated dogs expressing an agglutinating red cell antibody. Further studies with this drug indicated that both cytopenia [57,60] and suppression of bone marrow stem cell activity appear to be antibody mediated [61].

9.3.3   Immunostimulation

A variety of drugs as well as environmental chemicals have been shown to have immunostimulatory or sensitizing effects on the immune system, and these effects are well documented in humans exposed to drugs [8]. The drug or metabolite can act as a hapten and covalently bind to a protein or other cellular constituent of the host to appear foreign and become antigenic. Haptens are low-molecular-weight substances that are not in themselves immunogenic but will induce an immune response if conjugated with nucleophilic groups on proteins or other macromolecular carriers. In both allergy and autoimmunity, the immune system is stimulated or sensitized by the drug conjugate to produce specific pathological responses. An allergic hypersensitivity reaction may vary from one that results in an immediate anaphylactic response to one that produces a delayed hypersensitivity reaction or immune complex reaction.

Table 9.7 Drugs That Produce Immunostimulation

Drug	Type of response
Antibiotics	*Hypersensitivity*
Cephalosporins	Anaphylaxis, urticaria, rash, granulocytopenia
Chloramphenicol	Rash, dermatitis, urticaria
Neomycin	Dermal exposure rash, dermatitis
Sulfathiazole	Rash, dermatitis, urticaria
Spiramycin	Rash, dermatitis, urticaria
Quinolones	Photosensitivity
Tetracyclines	Photosensitivity, anaphylaxis, asthma, dermatitis
Others	
Allopurinol	Rash, urticaria, fever, eosinophilia
Avridine	Delayed-type hypersensitivity; increases NK cells, T cells,
Isoprinosine	IL-1, and IL-2
Indomethacin	Delayed-type hypersensitivity; increases T-lymphocytes
Quinidine	Rash, urticaria, asthma, granulocytopenia
Salicylates	Fever, anaphylaxis, asthma
	Rash, urticaria
	Autoimmunity
Amiodarone	Thyroiditis
Captopril	Autoimmune hemolytic anemia, pemphigus,
Chlorpromazine	granulocytopenia
Halothane	Granulocytopenia
Hydralizine	Autoimmune chronic active hepatitis
	Autoimmune hemolytic anemia, drug-induced SLE, myasthenia gravis, pemphigus, glomerulonephritis, Goodpasture's disease
Methyldopa	Autoimmune hemolytic anemia, leukopenia, drug-induced
Nitrofurantoin	SLE, pemphigus
D-Penicillamine	Peripheral neuritis
	Autoimmunity; drug-induced SLE, myasthenia gravis, pemphigus, glomerulonephritis, Goodpasture's disease
Propranolol	Autoimmunity
Procainamide	Autoimmunity, drug-induced SLE, rash, vasculitis, myalgias
Pyrithioxine	Pemphigus
Antibiotics	*Hypersensitivity and autoimmunity*
Isoniazid	Rash, dermatitis, vasculitis, arthritis, drug-induced SLE
Penicillins	Anaphylaxis, dermatitis; vasculitis, serum sickness, hemolytic anemia
Sulfonamides	Dermatitis, photosensitivity; pemphigus, hemolytic anemia, serum sickness, drug-induced SLE

Table 9.7 (Continued) Drugs That Produce Immunostimulation

Drug	Type of response
Others	
Acetazolamide	Rash, fever, autoimmunity
Lithium	Dermatitis; autoimmune thyroiditis, vasculitis
Thiazides	Hypersensitivity, photosensitivity; autoimmunity (diabetes)
Phenytoin	Rash; drug-induced SLE, hepatitis

Allergic hypersensitivity reactions result in a heightened sensitivity to nonself antigens, whereas autoimmunity results in an altered response to self antigens. Unlike immunosuppression, which nonspecifically affects all individuals in a dose-related manner, both allergy and autoimmunity have a genetic component that creates susceptibility in those individuals with a genetic predisposition. Susceptible individuals, once sensitized, can respond to even minute quantities of the antigen. Several examples of drugs that can stimulate the immune system are presented in Table 9.7.

9.3.3.1 Hypersensitivity

The four types of hypersensitivity reactions as classified by Coombs and Gell [59] are outlined in Table 9.8. The first three types are immediate antibody-mediated reactions, whereas the fourth type is a cellular-mediated delayed-type response that may require 1–2 days to occur after a secondary exposure. Type I reactions are characterized by an anaphylaxis response to a variety of compounds, including proteinaceous materials and pharmaceuticals such as penicillin. Various target organs may be involved, depending on the route of exposure. For example, the gastrointestinal tract is usually involved with food allergies, the respiratory system with inhaled allergens, the skin with dermal exposure, and smooth muscle vasculature with systemic exposure. The type of response elicited often depends on the site of exposure and includes dermatitis and urticaria (dermal), rhinitis and asthma (inhalation), increased gastrointestinal emptying (ingestion), and systemic anaphylactic shock (parenteral).

Type I Hypersensitivity. During an initial exposure, IgE antibodies are produced and bind to the cell surface of mast cells and basophils. Upon subsequent exposures to the antigen, reaginic IgE antibodies bound to the surface of target cells at the F_c region (mast cells and basophils) become cross-linked (at the F_{ab} regions) by the antigen. Cross-linking causes distortion of the cell surface and IgE molecule, which in turn activates a series of enzymatic reactions, ultimately leading to degranulation of the mast cells and basophils. These granules contain a variety of pharmacological substances (Table 9.9), such as histamines, serotonins, prostaglandins, bradykinins, and leukotrienes (SRS-A and ECF-A). Upon subsequent challenge exposures,

Table 9.8 Types of Hypersensitivity Responses [60]

Type and designation[a]	Agents: clinical manifestations	Components	Effects	Mechanism
I, Immediate (reaginic)	Food additives (GI allergies; anaphylactic) Penicillin: uticaria and dermatitis	Mast cells; IgE	Anaphylaxis, asthma, urticaria, rhinitis, dermatitis	IgE binds to mast cells to stimulate release of humoral factors
II, Cytotoxic	Cephalosporine: hemolytic anemia Quinidine: thrombocyto-penia	IgG, IgM	Hemolytic anemia, Goodpasture's disease	IgG and IgM bind to cells (e.g., red blood cells), fix complement (opsinization), then lyse cells
III, Immune complex (arthus)	Methicillin: chronic glomerulonephritis	Antigen–antibody complexes (Ag–Ab)	SLE, rheumatoid arthritis, glomerular nephritis, serum sickness, vasculitis	Ag–Ab complexes deposit in tissues and may fix complement
IV, Delayed hypersensitivity	Penicillin: contact dermatitis	T_D cells; macrophages	Contact dermatitis, tuberculosis	Sensitized T cells induce a delayed-hypersensitivity response upon challenge

Table 9.9 Proteins and Soluble Mediators Involved in Hypersensitivity [17]

Factor	Origin	Characteristics/functions
Histamine	Mast cells, basophils	Contraction of smooth muscle; increases vascular permeability
Serotonin	Mast cells, basophils	Contraction of smooth muscle; leukotriene
SRS-A	Lung tissue	(Slow-reacting substance of anaphylaxis); contraction of smooth muscle; acidic polypeptide
ECF-A	Mast cells	(Eosinophilic chemotactic factor of anaphylaxis); attracts eosinophils; small peptide
Prostaglandins	Various tissues	Modifies release of histamine and serotonin from mast cells and basophils

these factors are responsible for eliciting an allergic reaction through vasodilation and increased vascular permeability. The nasal passages contain both mast cells and plasma cells that secrete IgE antibodies. Allergic responses localized in the nasal mucosa result in dilation of the local blood vessels, tissue swelling, mucus secretion, and sneezing. Reactions localized in the respiratory tract, also rich in mast cells and IgE, result in an allergic asthma response. This condition is triggered by the release of histamine and SRS-A, which induce constriction of the bronchi and alveoli, pulmonary edema, and mucous secretions that block the bronchi and alveoli, together resulting in severe difficulty in breathing. In the case of a challenge dose of a drug administered systemically, the reactive patient may have difficulty breathing within minutes of exposure and may experience convulsions, vomiting, and low blood pressure. The effects of anaphylactic shock and respiratory distress, if severe, may ultimately result in death.

Antibiotics containing β-lactam structures, such as penicillin and cephalosporins, are the most commonly occurring inducers of anaphylactic shock and drug hypersensitivity in general. Other hypersensitivity reactions may include urticarial rash, fever, bronchospasm, serum sickness, and vasculitis with reported incidences of all types varying from 0.7% to 10% [62] and the incidence of anaphylactoid reactions varying from 0.04% to 0.2%. When the β-lactam ring is opened during metabolism, the penicilloyl moiety can form covalent conjugates with nucleophilic sites on proteins. The penicilloyl conjugates can then act as haptens to form the determinants for antibody induction. Although most patients who have received penicillin produce antibodies against the metabolite benzylpenicilloyl, only a fraction experience allergic reaction [63], which suggests a genetic component to susceptibility.

Type II Hypersensitivity. Type II cytolytic reactions are mediated by IgG and IgM antibodies that can fix complement, opsonize particles, or induce antibody-dependent cellular cytolysis reactions. Erythrocytes, lymphocytes, and platelets of the circulatory system are the major target cells that interact with the cytolytic antibodies causing depletion of these cells. Hemolytic anemia (penicillin, methyldopa), leukopenia, thrombocytopenia (quinidine), and/or granulocytopenia (sulfonamide) may result. Type II reactions involving the lungs and kidneys occur through the development of antibodies (autoantibodies) to the basement membranes in the alveoli or glomeruli, respectively. Prolonged damage may result in Goodpasture's disease, an autoimmune disease characterized by pulmonary hemorrhage and glomerulonephritis. Several other autoimmune-type diseases have been associated with extended treatments with D-penicillamine and other pharmaceuticals. Various types of autoimmune responses and examples of drug-induced autoimmunity are discussed in further detail later in this section.

Type III Hypersensitivity. Type III reactions (arthus) are characterized as an immediate hypersensitivity reaction initiated by antigen–antibody complexes that form freely in the plasma instead of at the cell surface. Regardless of whether the antigens are self or foreign, complexes mediated by IgG can form and settle into the tissue compartments of the host. These complexes can then fix complement and release C3a and C5a fragments that are chemotactic for phagocytic cells. Polymorphonuclear leukocytes are then attracted to the site, where they phagocytize the complexes and release hydrolytic enzymes into the tissues. Additional damage can be caused by binding to and activating platelets and basophils, which, in the end, results in localized necrosis, hemorrhage, and increased permeability of local blood vessels. These reactions commonly target the kidney, resulting in glomerulonephritis through the deposition of the complexes in the glomeruli.

Some antibiotics (β-lactam) have been reported to produce glomerular nephritis in humans that has been attributed to circulating immune complexes. These complexes have also been observed in preclinical toxicology studies with baboons treated with a β-lactam antibiotic, prior to the appearance of any biochemical or clinical changes [17]. In addition, immunoglobulin complexes have been observed in rats treated with gold, and autologus immune complex nephritis has been observed in guinea pigs [64]. Similar evidence of immunomediated nephrotoxicity has been reported in rheumatoid arthritis patients who were administered long-term treatments with gold compounds; proteinuria has been observed in approximately 10% of these patients.

Other target organs, such as the skin with lupus, the joints with rheumatoid arthritis, and the lungs with pneumonitis, may be affected. The deposition of antigen–antibody complexes through the circulatory system results in a syndrome referred to as *serum sickness*, which was quite prevalent prior to 1940 [65], when serum therapy for diphtheria was commonly used. Serum sickness occurs when the serum itself becomes antigenic as a side effect from passive immunization with heterologous antiserum produced from various sources of farm animals. The antitoxin for diphtheria was produced in a horse and administered to humans as multiple injections of passive antibody. As a consequence, these people often became sensitized to the horse serum and developed a severe form of arthritis and glomerulonephritis caused by deposition of antigen–antibody complexes. Clinical symptoms of serum sickness present as urticarial skin eruptions, arthralgia or arthritis, lymphadenopathy, and fever. Drugs such as sulfonamides, penicillin, and iodides can induce a similar type of reaction. Although uncommon today, transplant patients receiving immunosuppressive therapy with heterologous antilymphocyte serum or globulins may also exhibit serum sickness.

Type IV Delayed-Type Hypersensitivity (DTH). Delayed-type hypersensitivity reactions are T-cell mediated with no involvement of antibodies. However, these reactions are controlled through accessory cells, suppressor T cells, and monokine-secreting macrophages, which regulate the proliferation and differentiation of T cells. The most frequent form of DTH manifests itself as contact dermatitis. The drug or metabolite binds to a protein in the skin or the Langerhans cell membrane (class II MHC molecules) where it is recognized as an antigen and triggers cell proliferation. After a sufficient period of time for migration of the antigen and clonal expansion (latency period), a subsequent exposure will elicit a dermatitis reaction. A 24–48 h delay often occurs between the time of exposure and onset of symptoms to allow time for infiltration of lymphocytes to the site of exposure. The T cells (CD4$^+$) that react with the antigen are activated and release lymphokines that are chemotactic for monocytes and macrophages. Although these cells infiltrate to the site via the circulatory vessels, an intact lymphatic drainage system from the site is necessary since the reaction is initiated in drainage lymph nodes proximal to the site [17]. The release (degranulation) of enzymes and histamines from the macrophages may then result in tissue damage. Clinical symptoms of local dermal reactions may include a rash (not limited to sites of exposure) and itching and/or burning sensations. Erythema is generally observed in the area around the site, which may become thickened and hard to the touch. In severe cases, necrosis may appear in the center of the site followed by desquamation during the healing process. The immune-enhancing drugs isoprinosine and avridine have been shown to induce a delayed-type hypersensitivity reaction in rats [66].

A second form of delayed-type hypersensitivity response is similar to that of contact dermatitis in that macrophages are the primary effector cells responsible for stimulating CD4$^+$ T cells; however, this response is not necessarily localized to the epidermis. A classical example of this type of response is demonstrated by the tuberculin diagnostic tests. To determine if an individual has been exposed to tuberculosis, a small amount of fluid from tubercle bacilli cultures is injected subcutaneously. The development of induration after 48 h at the site of injection is diagnostic of prior exposure.

Shock, similar to that of anaphylaxis, may occur as a third form of a delayed systemic hypersensitivity response. However, unlike anaphylaxis, IgE antibodies are not involved. This type of response may occur 5–8 h after systemic exposure and can result in fatality within 24 h following intravenous or intraperitoneal injection.

A fourth form of delayed hypersensitivity results in the formation of granulomas. If the antigen is allowed to persist unchecked, macrophages and fibroblasts are recruited to the site to proliferate, produce collagen, and effectively "wall off" the antigen. A granuloma requires a minimum of 1 to 2 weeks to form.

9.3.3.2 *Photosensitization*

Regardless of the route of exposure, some haptens (photoantigens) that are absorbed locally into the skin, or reach the skin through systemic absorption, can be photoactivated by ultraviolet (UV) light between 320–400 nm. Once activated, the hapten can bind to the dermal receptors to initiate sensitization (photoallergy). Subsequent exposures to the hapten in the presence of UV light can result in a hypersensitivity response. Clinical symptoms of photoallergy may occur within minutes (immediate hypersensitivity) of exposure to sunlight, or 24 h or more after exposure (DTH). Symptoms may range from acute urticarial reactions to eczematous or papular lesions. Although both phototoxic and photoallergic reactions require the compound to be exposed to sunlight to elicit a response, their mechanisms of action are quite different. Since photosensitization is an immune-mediated condition, repeated exposures with a latency period between the initial exposure and subsequent exposures is required, the response is not dose related (small amounts can produce a response when sensitized), and not all individuals exposed to the compound will necessarily respond (genetic component to susceptibility). Although both conditions can present similar symptoms (erythema), phototoxicity is limited mainly to erythema, whereas photoallergy can result in erythema, edema, and dermatitis as described above.

Several drug classes, including tetracycline, sulfonamide, and quinolone antibiotics, as well as chlorothiazide, chlorpromazine, and amiodarone hydrochloride, have been shown to be photoantigens. Photosensitivity may persist even after withdrawal of the drug, as has been observed with the antiarrhythmic drug amiodarone hydrochloride, since it is lipophilic and can be stored for extended periods in the body fat [67]. In addition, it is quite common for cross-reactions to occur between structurally related drugs of the same class.

9.3.3.3 Autoimmunity

In autoimmunity, as with hypersensitivity, the immune system is stimulated by specific responses that are pathogenic, and both tend to have a genetic component that predisposes some individuals more than others. However, as is the case with hypersensitivity, the adverse immune response of drug-induced autoimmunity is not restricted to the drug itself but also involves a response to self antigens.

Autoimmune responses directed against normal components of the body may consist of antibody-driven humoral responses and/or cell-mediated, delayed-type hypersensitivity responses. T cells can react directly against specific target organs, or B cells can secrete autoantibodies that target self. Autoimmunity may occur spontaneously as the result of a loss of regulatory controls that initiate or suppress normal immunity, causing the immune system to produce lymphocytes reactive against its own cells and macromolecules such as DNA, RNA, or erythrocytes.

Although autoantibodies are often associated with autoimmune reactions, they are not necessarily indicative of autoimmunity [68]. Antinuclear antibodies can occur normally with aging in some healthy women without autoimmune disease, and all individuals have B cells with the potential of reacting with self antigens through Ig receptors [69]. The presence of an antibody titer to a particular immunogen indicates that haptenization of serum albumin has occurred as part of a normal immune response. However, if cells are stimulated to proliferate and secrete autoantibodies directed against a specific cell or cellular component, a pathological response may result. The tissue damage associated with autoimmune disease is usually a consequence of type II or III hypersensitivity reactions that result in the deposition of antibody–antigen complexes.

Several diseases have been associated with the production of auto-antibodies against various tissues. For example, an autoimmune form of hemolytic anemia can occur if the antibodies are directed against erythrocytes. Similarly, antibodies that react with acetylcholine receptors may cause myasthenia gravis; those directed against glomerular basement membranes may cause Goodpasture's syndrome; and those that target the liver may cause hepatitis. Other forms of organ-specific autoimmunity include autoimmune thyroiditis (as seen with amiodarone) and juvenile diabetes mellitus, which can result from autoantibodies directed against the tissue-specific antigens thyroglobulin and cytoplasmic components of pancreatic islet cells, respectively. In contrast, systemic autoimmune diseases may occur if the autoantibodies are directed against an antigen that is ubiquitous throughout the body, such as DNA or RNA. For example, systemic lupus erythematosus (SLE) occurs as the result of autoimmunity to nuclear antigens that form immune complexes in the walls of blood vessels and basement membranes of tissues throughout the body.

The etiology of drug-induced autoimmunity is not well established and is confounded by factors such as age, sex, and nutritional state, as well as genetic influences on pharmacological and immune susceptibility. Unlike idiopathic autoimmunity, which is progressive or characterized by an alternating series of relapses and remissions, drug-induced autoimmunity is thought to subside after the drug is discontinued. However, this is not certain since a major determining factor for diagnosis of a drug-related disorder is dependent on the observation of remission upon withdrawal of the drug [70].

One possible mechanism for xenobiotic-induced autoimmunity involves xenobiotic binding to autologus molecules, which then appear foreign to the immunosurveillance system. If a self antigen is chemically altered, a specific T helper (T_h) cell may see it as foreign and react to the altered antigenic determinant portion, allowing an autoreactive B cell to react to the unaltered hapten. This interaction results in a carrier–hapten bridge between the specific T_h and autoreactive B cell, bringing them together for subsequent production of autoantibodies specific to the self antigen that was chemically altered [71]. Conversely, a xenobiotic may alter B cells directly, including those that are autoreactive. Thus, the altered B cells may react to self antigens independent from T_h-cell recognition and in a nontissue-specific manner.

Another possible mechanism is that the xenobiotic may stimulate nonspecific mitogenicity of B cells. This could result in a polyclonal activation of B cells with subsequent production of autoantibodies. Alternatively, the xenobiotic may stimulate mitogenicity of T cells that recognize self, which in turn activate B-cell production of antibodies in response to self molecules. There is also evidence to suggest that anti-DNA autoantibodies may originate from somatic mutations in lymphocyte precursors with antibacterial or antiviral specificity. For example, a single amino acid substitution resulting from a mutation in a monoclonal antibody to polyphorylcholine was shown to result in a loss of the original specificity and an acquisition of DNA reactivity similar to that observed for anti-DNA antibodies in SLE [72].

The mechanisms of autoimmunity may also entail interaction with MHC structures determined by the human leukocyte antigen (HLA) alleles. Individuals carrying certain HLA alleles have been shown to be predisposed to certain autoimmune diseases, which may account in part for the genetic variability of autoimmunity. In addition, metabolites of a particular drug may vary between individuals to confound the development of drug-induced autoimmunity. Dendritic cells, such as the Langerhans cells of the skin and B lymphocytes that function to present antigens to T_h cells, express class-II MHC structures. Although the exact involvement of these MHC structures is unknown, Gleichmann et al. [7] have theorized that self antigens rendered foreign by drugs such as D-penicillamine may be presented to T_h cells by MHC class-II structures.

An alternate hypothesis is that the drug or a metabolite may alter MHC class-II structures on B cells, making them appear foreign to T_h cells.

A number of different drugs have been shown to induce autoimmunity in susceptible individuals. A syndrome similar to that of SLE was described in a patient administered sulfadiazine in 1945. Sulfonamides were one of the first classes of drugs identified to induce an autoimmune response, while to date, more than 40 other drugs have been associated with a similar syndrome [72].

Autoantibodies to red blood cells and autoimmune hemolytic anemia have been observed in patients treated with numerous drugs, including procainamide, chlorpropamide, captopril, cefalexin, penicillin, and methyldopa [73,74]. Hydralazine- and procainamide-induced autoantibodies may also result in SLE. Approximately 20% of patients administered methyldopa for several weeks for the treatment of essential hypertension developed a dose-related titer and incidence of autoantibodies to erythrocytes, 1% of which presented with hemolytic anemia. Methlydopa does not appear to act as a hapten but appears to act by modifying erythrocyte surface antigens. IgG autoantibodies then develop against the modified erythrocytes.

D-penicillamine is used to treat patients with rheumatoid arthritis, to reduce excess cystine excretion in patients with cystinurias, and as a chelating agent for copper in patients with Wilson's disease. D-penicillamine can cause multiple forms of autoimmunity including SLE, myasthenia gravis, pemphigus, and autoimmune thyroiditis. This drug is thought to act as immunomodulator in patients by initiating or even potentiating anti-DNA antibody synthesis [75]. The highly reactive thiol group may react with various receptors and biological macromolecules to induce autoantibodies. Long-term (many months) treatment has been shown to induce autoimmunity resulting in myasthenia gravis in 0.5% of patients [70] and SLE in approximately 2% of patients as exhibited by varying degrees of joint pain, synovitis, myalgia, malaise, rash, nephritis, pleurisy, and neurological effects. In patients exhibiting myasthenia gravis, D-penicillamine may act to alter the acetylcholine receptors. Autoantibodies to acetylcholine receptors have been detected in these patients and have been shown to decrease gradually after drug withdrawal concomitant with reversibility of the clinical syndrome. However, myasthenia gravis may persist for long periods of time after D-penicillamine therapy has ceased.

Although rare, cases of renal lupus syndrome and pemphigus blisters have also been reported as a consequence of D-penicillamine-induced immune complexes [70,76], as well as with other drugs. With renal lupus syndrome, secondary glomerulonephritis may result if granular IgG antibodies are produced and deposited on the basement membranes. In patients with pemphigus blisters, autoantibodies to the intercellular substance of the skin have been recovered from the sera, and dermal biopsies have demonstrated intracellular deposits or immunoglobulin deposits on the basement membranes. Pemphigus has also been observed in patients treated with sulfhydryl compounds such as captopril and pyrithioxine [70].

Some metals that are used therapeutically have also been shown to induce autoimmune responses. Gold salts used to treat arthritis may induce formation of antiglomerular basement membrane antibodies, which may lead to glomerulonephritis similar to that seen in Goodpasture's disease (see type II hypersensitivity). Since gold is not observed at the site of the lesions [77] it has been hypothesized that the metal elicits an antiself response. Lithium, used to treat manic-depression, is thought to induce autoantibodies against thyroglobulin, which in some patients results in hypothyroidism. In studies with rats, levels of antibodies to thyroglobulin were shown to increase significantly in lithium-treated rats compared to controls immediately after immunization with thyroglobulin; however, rats that were not immunized with thyroglobulin did not produce circulation antithyroglobulin antibodies upon receiving lithium, and there was no effect of lithium on lymphocytic infiltration of the thyroid in either group [78].

Some drugs such as penicillin have been shown to induce autoimmunity as well as anaphylaxis [7]. The carbonyl of the β-lactam ring of penicillin can form a covalent penicilloyl conjugate with nucleophilic sites on proteins, particularly the amino groups of lysine residues. This conjugate, which acts as the major immunogenic determinant, may become biotransformed to other isomeric forms of clinical relevance [79].

A genetic predisposition to drug-induced development of SLE has been shown to occur in some individuals treated with the drugs hydralazine, isoniazid, procainamide, and sulphamethazine. A polymorphism, which is known to exist for the genes responsible for expression of hepatic N-acetyl transferase enzymes, determines the rate of acetylation of these drugs to regulate the rate of drug inactivation. Individuals who are relatively slow acetylators of these drugs are more likely to develop antinuclear antibodies and are at a higher risk for developing SLE [80]. Other predisposing factors, such as HLA phenotype (HLA-DR4 and/or C4 allele), may also play a genetic role in determining susceptibility to hydralazine-induced SLE [81].

In addition, silicone-containing medical devices, particularly breast prostheses, have been reported to cause serum-sickness-like reactions, scleroderma-like lesions, and an SLE-like disease termed *human adjuvant disease* [11,23]. Some patients may also present with granulomas and autoantibodies. Human adjuvant disease is a connective tissue or autoimmune disease similar to that of adjuvant arthritis in rats and rheumatoid arthritis in humans. Autoimmune disease–like symptoms usually develop 2–5 y after implantation in a small percentage of people that receive implants, which may indicate that there is a genetic predisposition similar to that for

SLE. An early hypothesis is that the prosthesis or injected silicone plays an adjuvant role by enhancing the immune response through increased macrophage and T-cell helper function. There is currently controversy as to whether silicone, as a foreign body, induces a nonspecific inflammation reaction, a specific cell-mediated immunological reaction, or no reaction at all. However, there is strong support to indicate that silicone microparticles can act as haptens to produce a delayed hypersensitivity reaction in a genetically susceptible population of people.

References

1. ICH, S8 *Investigational Evaluation of Investigational New Drugs*, 2005.
2. ICH, S7A *Safety Pharmacology Studies for Human Pharmacology*, July 2001.
3. Patterson, R., et al., Drug allergies and protocols for management of drug allergies, *NER Allergy*, 1986, Proc. 7:325–342.
4. Hastings, K.L., Pre-clinical methods for detecting the hypersensitivity potential of pharmaceuticals: regulatory consideration, *Toxicology*, 2001, 158:85–89.
5. Hutchings, P., Nador, D., and Cooke, A., Effects of low doses of cyclophosphamide and low doses of irradiation on the regulation of induced erythrocyte autoantibodies in mice, *Immunol.*, 1985, 54:97–104.
6. Koller, L.D., Immunotoxicology today, *Toxicol. Path.*, 1987, 15:346–351.
7. Gleichmann, E., Kimber, I., and Purchase, I.F.H., Immunotoxicology: suppressive and stimulatory effects of drugs and environmental chemicals on the immune system, *Arch. Toxicol.*, 1989, 63:257–273.
8. DeSwarte, R.D., Drug allergy: an overview, *Clin. Rev. Allergy*, 1986, 4:143–169.
9. Choquet-Kastylevsky, G., Vial, T., and Descotes, J., Drug allergy diagnosis in humans: possibilities and pitfalls, *Toxicology*, 2001, 158:1–10.
10. Pieters, R., The popliteal lymph node assay: a tool for predicting drug allergies, *Toxicology*, 2001, 158:65–69.
11. Kumagai, Y., et al., Clinical spectrum of connective tissue disease after cosmetic surgery: observations of 18 patients and a review of the Japanese literature, *Arthr. Rheum.*, 1984, 27:1–12.
12. DeWeck, A.L., Immunopathological mechanisms and clinical aspects of allergic reactions to drugs. In: *Handbook of Experimental Pharmacology: Allergic Reactions to Drugs* (deWeck, A.L., and Bundgaard, H., eds.), New York: Springer-Verlag, 1983, pp. 75–133.
13. Atkinson, T.P., and Kaliner, M.A., Anaphylaxis medical elimination, *North America*, 1992, 76:841–855.
14. Van der Klann, N.M., Wilson, J.H.P, and Stricker, B.H., Drug assisted anaphylaxis: 20 years reporting in the southern lands, *Elim. Erop. Allergy*, 1996, 26:1355–1363.
15. Middleton, E.P., et al., *Allergy: Principles and Practice*, 4th ed., Baltimore: Crosby, 2002.
16. Luster, M.L., Blank, J.A., and Dean, J.H., Molecular and cellular basis of chemically induced immunotoxicity, *Ann. Rev. Pharmacol. Toxicol.*, 1987, 27:23–49.
17. Descotes, G., and Mazue, G., Immunotoxicology, *Advances in Veterinary Science and Comparative Medicine*, 1987, 31:95–119.

18. Sarlos, K., and Clark, E.D., *Evaluating Chemicals as Respiratory Allergens: Using the Tieh Approach for Risk Assessment, Methods in Immunotoxicology*, vol. 2 (Burleson, G.R., Dean, J.H., and Munson, A.E., eds.), New York: Wiley, 1995, pp. 411–421.

19. *CDER Guidance for Industry: Immunotoxicology Evaluation of Investigational New Drugs*. U.S. Department of Health and Human Services, 2001.

20. Ader, R., and Cohen, N., Psychoneuroimmunology: conditioning and stress, *Ann. Rev. Psychol.* 1993, 44:53–85.

21. Male, D., Champion, B., and Cooke, A., *Advanced Immunology*, Philadelphia: J.B. Lippincott, 1982.

22. Roitt, I.M., Brostoff, J., and Male, D.K., *Immunology*, St. Louis: C.V. Mosby, 1985.

23. Guillaume, J.C., Roujeau, J.C., and Touraine, R., Lupus systémique après protheses mammaires, *Ann. Derm. Verner.*, 1984, 111:703–704.

24. Yoshida, S., Golub, M.S., and Gershwin, M.E., Immunological aspects of toxicology: premises not promises. *Reg. Toxicol. Pharm.*, 1989, 9:56–80.

25. Penn, I., Development of cancer as a complication of clinical transplantation, *Transplant Proc.*, 1977, 9:1121–1127.

26. Burnet, F.M., The concept of immunological surveillance, *Progr. Exper. Tumor. Res.*, 1970, 13:1–27.

27. Merluzzi, V.J., Comparison of murine lymphokine, activated killer cells, natural killer cells, and cytotoxic T lymphocytes. *Cell Immunol.*, 1985, 95:95–104.

28. Volkman, A., *Mononuclear Phagocyte Function*, New York: Marcel Dekker, 1984.

29. Bakke, O.M., Wardell, W.M., and Lasagna, L., Drug discontinuations in the United Kingdom and United States, 1964–1983: Issues of safety, *Clin. Pharmacol. Therapy*, 1984, 35:559–567.

30. Hunter, T., et al., Azathioprine in rheumatoid arthritis: a long-term follow-up study. *Arthr. Rheum.*, 1975, 8(I):15–20.

31. Hadden, J.W., Cornaglia-Ferraris, P., and Coffey, R.G., Purine analogs as immunomodulators. In: *Progress in Immunology IV* (Yamamura, Y., and Tada, T., eds.), London: Academic Press, 1984, pp. 1393–1407.

32. Calabresi, P., and Chabner, B.A., Antineoplastic agents. In: *The Pharmacological Basis of Therapeutics* (Goodman, A.G., et al., eds.), New York: Pergamon Press, 1990, pp. 1209–1263.

33. Spreafico, F., and Anaclerio, A., Immunosuppressive agents. In: *Immunopharmacology 3*, (Hadden, J., Coffey, R., and Spreafico, R., eds.), New York: Plenum Medical Book Company, 1977, pp. 245–278.

34. Alper, J.C., et al., Rationally designed combination chemotherapy for the treatment of patients with recalcitrant psoriasis, *J. Am. Acad. Dermatol.*, 1985, 13:567–577.

35. Barnett, M.J., et al., High-dose cytosine arabinoside in the initial treatment of acute leukemia, *Semin. Oncol.*, 1985, 12:133–138.

36. Elion, G.B., and Hitchings, G.H., Azathioprine. In: *Antineoplastic and Immunosuppressive Agents* (Sartorelli, A.C., and Johns, D.G., eds.), Berlin: Springer-Verlag, 1975, pp. 403–425.

37. Hollenberg, S.M., et al., Colocalization of DNA-binding and transcriptional activation functions in the human glucocorticoid receptor, *Cell*, 1987, 49:39–46.

38. Wallner, B.P., et al., Cloning and expression of human lipocortin, a phospholipase A2 inhibitor with potential anti-inflammatory activity, *Nature*, 1986, 320:77–80.

39. Kay, J.E., and Benzie, C.R., Rapid loss of sensitivity of mitogen-induced lymphocyte activation to inhibition by cyclosporin A, *Cell Immunol.*, 1984, 87:217–224.

40. Elliot, J.F., et al., Induction of interleukin 2 messenger RNA inhibited by cyclosporin A, *Science*, 1984, 226:1439–1441.

41. Herold, K.C., et al., Immunosuppressive effects of cyclosporin A on cloned T cells, *J. Immunol*, 1986, 136:1315–1321.

42. Kahan, B.D., and Bach, J.F., Proceedings of the Second International Congress on Cyclosporine. *Transplant. Proc.*, 1988, 20(Suppl.):11131.

43. Colvin, M., The alkylating agents. In: *Pharmacologic Principles of Cancer Treatment* (Chabner, B.A., ed.), Philadelphia: W.B. Saunders, 1982, p. 276–308.

44. Calabresi, P., and Parks, R., Antiproliferative agents and drugs used for immunosuppression. In: *Goodman and Gilman's The Pharmacological Basis of Therapeutics*, 7th ed. (Gilman, A.G., et al., eds.), New York: Macmillan Publishing, 1985, pp. 1247–1306.

45. Webb, D.R., and Winklestein, A., Immunosuppression, immunopotentiation, and anti-inflammatory drugs. In: *Basic and Clinical Immunology*, 4th ed. (Stites, D.P., et al., eds.), Los Altos, CA: Lange Medical, 1982, pp. 277–292.

46. Shand, F.L., Review/commentary: the immunopharmacology of cyclophosphamide, *Int. J. Immunopharm.*, 1979, 1:165–171.

47. Luster, M.L., et al., Estrogen immunosuppression is regulated through estrogenic responses in the thymus, *J. Immunol.*, 1984, 133:110–116.

48. Pung, O.J., et al., Influence of steroidal and nonsteroidal sex hormones on host resistance in the mouse: increased susceptibility to Listeria monocytogenes following exposure to estrogenic hormones, *Infect. Immun.*, 1984, 46:301–307.

49. Luster, M.L., Pfeifer, R.W., and Tucher, A.N., Influence of sex hormones on immunoregulation with specific reference to natural and synthetic estrogens. In: *Endocrine Toxicology* (McLachlin, J.A., Korach, K., and Thomas, J., eds.), New York: Raven, 1985, pp. 67–83.

50. Dean, J.H., et al., The effect of adult exposure to diethylstilbestrol in the mouse: alterations in tumor susceptibility and host resistance parameters, *J. Reticuloendothelial. Soc.*, 1980, 28:571–583.

51. Luster, M.L., et al., The effect of adult exposure to diethylstilbestrol in the mouse: alterations in immunological function, *J. Reticuloendothel. Soc.*, 1980, 28:561–569.

52. Haukaas, S.A., Hoisater, P.A., and Kalland, T., *In vitro* and *in vivo* effects of diethylstilbestrol and estramustine phosphate (Estracyte) on the mutagen responsiveness of human peripheral blood lymphocytes, *Prostate*, 1982, 3:405–414.

53. Ways, S.C., et al., Alterations in immune responsiveness in women exposed to diethylstilbestrol in utero, *Fertil. Steril.*, 1987, 48:193–197.

54. Pung, O.J., et al., Influence of estrogen on host resistance: increased susceptibility of mice to Listeria monocytogenes correlates with depressed production of interleukin 2, *Infect. Immun.*, 1985, 50:91–96.

55. Lawrence, D.A., Immunotoxicity of heavy metals. In: *Immunotoxicology and Immunopharmacology* (Dean, J.H., et al., eds.), New York: Raven, 1985, pp. 341–353.

56. Bloom, J.C., Thiem, P.A., and Morgan, D.G., The role of conventional pathology and toxicology in evaluating the immunotoxic potential of xenobiotics, *Toxicol. Path.*, 1987, 15:283–293.

57. Bloom, J.C., et al., Cephalosporin-induced immune cytopenia in the dog: demonstration of cell-associated antibodies, *Blood*, 1985, 66:1232.

58. Caspritz, G., and Hadden, J., The immunopharmacology of immunotoxicology and immunorestoration, *Toxicol. Path.*, 1987, 15:320–322.

59. Coombs, R.R.A., and Gell, P.G.H., Classification of allergic reactions responsible for clinical hypersensitivity and disease. In: *Clinical Aspects of Immunology* (Gell, P.G.H., Coombs, R.R.A., and Lachman, D.J., eds.), Oxford: Blackwell Scientific Publications, 1975, p. 761.

60. Bloom, J.C., et al., Gold-induced immune thrombocytopenia in the dog, *Vet. Pathol.*, 1985, 22:492–499.

61. Deldar, A., et al., Residual stem cell defects associated with cephalosporin therapy in dogs, *Blood*, 1985, 66:1202.

62. Idsøe, O., et al., Nature and extent of penicillin side-reactions, with particular reference to fatalities from anaphylactic shock, *Bull. WHO*, 1968, 38:159–188.

63. Garratty, G., and Petz, L.D., Drug-induced immune hemolytic anemia, *Am. J. Med.*, 1975, 58:398–407.

64. Ueda, S., et al., Autologous immune complex nephritis in gold injected guinea pigs, *Nippon Jinzo Gakkai Shi*, 1980, 22:1221–1230.

65. Clark, W.R., *The Experimental Foundations of Modern Immunology*, 2nd ed., New York: John Wiley & Sons, 1983, pp. 1–453.

66. Exon, J.H., et al., Immunotoxicology testing: an economical multiple assay approach, *Fund. Appl. Toxicol.*, 1986, 7:387–397.

67. Unkovic, J., et al., Poster, Annual Meeting of the American Society of Dermatology, 1984, Washington, D.C.

68. Russel, A.S., Drug-induced autoimmune disease, *Clin. Immun. Allergy*, 1981, 1:57.

69. Dighiero, G., et al., Murine hybridomas secreting natural monoclonal antibodies reacting with self antigens. *J. Immunol.*, 1983, 135:2267–2271.

70. Bigazzi, P.E., Autoimmunity induced by chemicals, *Clin. Toxicol.*, 1988, 26:125–126.

71. Weigle, W.O., Analysis of autoimmunity through experimental models of thyroiditis and allergic encephalomyelitis, *Adv. Immunol.*, 1980, 30:159–275.

72. Talal, N., Autoimmune mechanisms in patients and animal models, *Toxicol. Path.*, 1987, 15:272–275.

73. Logue, G.L., Boyd, A.E., and Rosse, W.F., Chlorpropamide-induced immune hemolytic anemia, *New Engl. J. Med.*, 1970, 283:900–904.

74. Kleinman, S., et al., Positive direct antiglobulin tests and immune hemolytic anemia in patients receiving procainamide, *New Engl. J. Med.*, 1984, 311:809–812.

75. Mach, P.S., Brouilhet, H., and Smor, B., d-penicillamine: a modulator of anti-DNA antibody production, *Clin. Exp. Immunol.*, 1986, 63:414–418.

76. Ntoso, K.A., et al., Penicillamine-induced rapidly progressive glomerulonephritis in patients with progressive systemic sclerosis: successful treatment of two patients and a review of the literature, *Amer. J. Kidney Dis.*, 1986, 8:159–163.

77. Druet, P., et al., Immunologically mediated glomerulonephritis induced by heavy metals, *Arch. Toxicol.*, 1982, 50:187–194.

78. Hassman, R.A., et al., The influence of lithium chloride on experimental autoimmune thyroid disease, *Clin Exp. Immunol.*, 1985, 61:49–57.
79. Batchelor, F.R., Dewdney, J.M., and Cazzard, D., Penicillin allergy: the formation of penicilloyl determinant. *Nature* (London), 1965, 206:362–364.
80. Perry, H.M., Tane, M., and Camody, S., Relationship of acetyl transferase activity to antinuclear antibodies and toxic symptoms in hypertensive patients treated with hydralazine, *J. Lab. Clin. Med.*, 1970, 76:114–125.
81. Spears, C.J., and Batchelor, J.R., Drug-induced autoimmune disease, *Adv. Nephrol.*, 1987, 16:219–230.
82. Golub, E.S., and Green, D.R., *Immunology: A Synthesis,* Sunderland, MA: Sinauer, 1991, pp. 1–744.
83. Gilman, A.G., et al., *The Pharmacological Basis of Therapeutics,* 8th ed., New York: Pergamon Press, 1990, pp. 1–1811.

Appendix A: Acronyms

510(k)	Section of the Federal Food, Drug, and Cosmetics act that allows for clearance of class II medical devices
ADME	Absorption distribution metabolism excretion
ADR	Adverse drug reaction
ANDA	Abbreviated new drug application
APD	Action potential duration
ARF	Acute renal failure
ATS	Academy of Toxicological Sciences
BBB	Blood-brain barrier
CCK	Cholecystokinin
CCTV	Closed circuit television
Cmax	Maximum concentration
CDER	Center for drug evaluation and research
CO	Cardiac output
CPMP	Committee for proprietary medicinal products
CNS	Central nervous system
CPG	Central pattern generator
CRO	Contract research organization
CSI	Cardiac safety index
CTCAE	Common terminology criteria for adverse events
CTD	Comparative toxicogenomics database
CVP	Central venous pressure
DA	Dopamine
DABT	Diplomate of the American Board of Toxicologists

DAP	Diastolic arterial blood pressure
DMPK	Dystrophia myotonica protein kinase
DTH	Delayed-type hypersensitivity
ECG/EKG	Electrocardiogram
ECHO	Echocardiography
EC10	Effective concentration in 10% of the population
EC90	Effective concentration in 90% of the population
ECS	Electroconvulsive stimulation
EDA	Exploratory data analysis
ED90	Dose of a therapeutic agent that eliminates 90% of the target pathogen
EEG	Electroencephalogram
EMA	European medicines agency
EMG	Electromyogram
EP	Evoked particles
EU	European Union
FBD	Functional bowel disorder
FDA	Food and Drug Administration
FIM	First exposure in man
FOB	Functional observational battery
GFR	Glomerular filtration rate
GH	Growth hormone
GLP	Good Laboratory Practice
HC	Heidelberg pH Capsule
hERG	Human *Ether-a-go-go*-related gene
IBS	Irritable bowl syndrome
ICH	International Conference on Harmonisation of Technical Requirements for Registration of Pharmaceuticals for Human Use
ICH S7A	ICH guideline covering selection and design of *in vitro* and *in vivo* studies with particular focus on respiratory, cardiovascular, and central nervous systems
ICHS7B	ICH guideline covering nonclinical evaluation of the potential for delayed ventricular depolarization (QT prolongation)
IDE	Investigational device exemption
IND	Investigational new drug
LOEL	Lowest observed effect level
LQTS	Long QT syndrome
LVP	Lateral ventricular pressure
MAP	Mean arterial blood pressure
MiRP1	MinK-related peptide 1
MinK	Minimal potassium ion channel

MPO	Myeloperoxidase
MRI	Magnetic resonance imaging
MT	Metallothionein
NDA	New drug application
NIH	National Institute of Health
NMDA	N-methyl-D-aspartic acid
NMR	Nuclear magnetic resonance
PAH	Para-aminohippuric acid
PAP	Pulmonary arterial pressure
PDE	Phospho-Diesterase
PG	Prostaglandin
PLA	Product License Application
PMA	Product Marketing Application
PTH	Parathyroid hormone
QRS	Three of the graphical deflections seen in an electrocardiogram, corresponding to the depolarization of the left and right ventricles; generally the most visible features of the ECG
QSAR	Quantitative structure-activity relationship
QT	Interval between beginning of Q wave and end of T wave in the heart's electrical cycle
QTc	Corrected QT interval
RF	Radio Frequency
RPF	Renal Plasma Flow
RR	Relative rate
SAP	Systolic arterial blood pressure
SHR	Spontaneously hypertensive rats
SLE	Systemic lupus erythematosus
SOT	Society of Toxicology
TdP	*Torsades de pointes*

Appendix B:
Laboratories conducting safety pharmacology testing

Lab	Location	Web site	Phone number	Services
Calvert	Olyphant, PA	www.calvertlabs.com	(570) 586-2411	Radioactivity Telemetry, Mouse Tumor Model
CIT	Cedex, France	www.citox.com	+33 2 32 292626	Clinical
Covance	Madison, WI Harrogate, U.K. Muenster, FRG	www.covance.com	(888)COVANCE	Clinical, Consulting
Huntingdon	East Millstone, NJ Cambridgeshire, U.K.	www.huntingdon.com	(732) 873-2550 44 (0) 1480892000	Radioactivity, Discovery Support
ITR	Montreal, QU	www.ITRlab.com	(514) 457-7400	
Laboratory of Pharmacology and Toxicology (LPT)	Hamburg, GE	www.LPT-pharm-tox.de	49 40-70-20-20	
MPI	Mattawan, MI	www.MPIresearch.com	(269) 668-3336	Safety Pharm, Transgenic mouse CA, Radioactivity
Porsolt & Partners	Boulougne- Billancourt, FR	www.porsolt.com	+33146109990	
Ricera	Concord, OH	www.ricera.com	(888) 763-4797	Synthesis, Product Development, Radioactivity
Shin Nippon Biomedical Laboratories (SNBL)	Tokyo, Japan	www.snbl.com	+81 355655001	Clinical, Irritation
SNBL USA, Ltd.	Everett, WA	www.snblusa.com	(425) 407-0121	
TNO	Zeist, The Netherlands	www.voeding.tno.nl	+31 30 694 4806	
WIL Research Labs	Ashland, OH	www.wilresearch.com	(419) 289-8700	Formulation

Index